EXPLORING MATHEMATICS ON YOUR OWN

Exploring Mathematics on Your Own

DONOVAN A. JOHNSON
Professor of Mathematics Education
University of Minnesota

WILLIAM H. GLENN
Formerly Mathematics Supervisor
Pasadena City Schools, Pasadena, California

DOVER PUBLICATIONS, INC., NEW YORK

Published in Canada by General Publishing
Company, Ltd., 30 Lesmill Road, Don Mills,
Toronto, Ontario.
Published in the United Kingdom by Constable
and Company, Ltd., 10 Orange Street, London WC 2.

This Dover edition, first published in 1972, is an
unabridged republication of the work originally
published by Webster Publishing Company in 1960.
"Paul Bunyan versus the Conveyer Belt," by
William Hazlett Upson, first published in *Ford
Times*, copyright 1949 by Ford Motor Company, is
reprinted by permission of the copyright owners.

International Standard Book Number: 0-486-20383-2
Library of Congress Catalog Card Number: 72-81535

Manufactured in the United States of America
Dover Publications, Inc.
180 Varick Street
New York, N. Y. 10014

Contents

PART I

xxxxxx
xx
xxxxxx
xxx
xxxxxx
xxxx
xxxxxx
xxxxx

Understanding
xxxxxx
Numeration Systems
xxxxxx
xxxxxx
xxxxxx

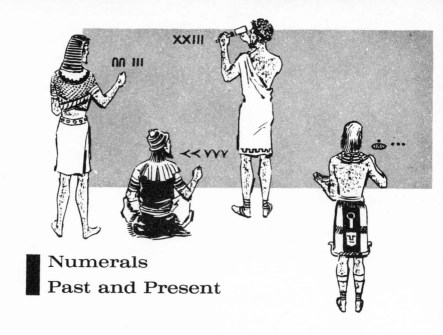

Numerals Past and Present

Numerals, Numbers, and Systems

Probably the greatest invention ever made by man was a number system. Although it is simple for us to use the symbol 23 to represent twenty-three objects, we should realize that it took man many centuries to build the system in which the figures 2 and 3, arranged in that order, do stand for twenty-three things.

The numbering system we use today was not the only one or the first one developed by man. Here are some of the ways that 23 has been written in the past:

Egyptian ∩∩|||

Babylonian << ΥΥΥ

Roman XXIII

Mayan ☼ •••

The amount represented by the symbols above is the same, but the symbols used are different. The symbols used to represent numbers are called *numerals*. A numeral, then, is a name or symbol used to stand for a number. When a certain set of symbols is used to represent numbers, a *numeration system* is formed. In a numeration system it is possible to use only a few symbols to represent many different numbers. This is done by arranging different combinations of the symbols and by giving importance to the order in which the symbols are written.

The numeration system we use is called the *Hindu-Arabic system*. It has this name because it was invented in India and brought to Europe by the Arabs. Most people believe that our system of numerals is a "natural" system, or the best system, or the only system we need. This is not true. In fact, in our modern world of atomic energy, space travel, and electronic computing machines we often find a need for different numeration systems. Some day someone may invent a numeration system far better than our present one — just as our system was far better than the Roman system.

If we consider other numeration systems, like the ones that we will describe in the following pages, we will have a better understanding of our numerals, how our computation processes work, and how new systems are being used in science.

Understanding Our Numeration System

Let us now take a close look at the numbers we use every day. Our numeration system has ten basic symbols: 0, 1, 2, 3, 4, 5, 6, 7, 8, 9. These symbols are called *digits*. We use combinations of our ten number symbols to write numerals for very large or very small numbers. This is possible because we assign different values to a digit, depending upon its position in a numeral, or its *place value*. We can illustrate this with a counting example. When we count in our system, we usually group by tens. In Figure 1, we have one group of ten figures, and three left over. Thus, these figures are counted to be 13. The numeral 13 means $(1 \times 10) + (3 \times 1)$. When we count, we usually work with "whole things" and thus we say that 1, 2, 3 . . . represent whole numbers.

Figure 1

Consider now the large group of x's in Figure 2. We can draw lines around groups of ten until fewer than ten remain. Then we can draw a line around a group of ten tens or a hundred.

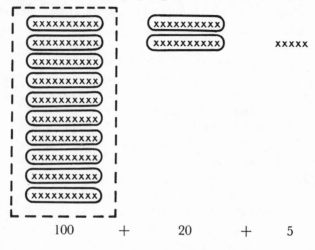

$$100 \quad + \quad 20 \quad + \quad 5$$

Figure 2

The number of x's can be represented by the numeral 125, which means $(1 \times 10 \times 10) + (2 \times 10) + (5 \times 1)$.

When we do addition problems, we often have to "regroup" numbers to form groups of ten. We often speak of this regrouping as "carrying" to the next digit place.

$$
\begin{aligned}
18 + 5 &= (10 + 8) + (2 + 3) \\
&= 10 + (8 + 2) + 3 \\
&= 10 + 10 + 3 \\
&= 23
\end{aligned}
$$

We have seen that our numeral system uses ten basic symbols and that the size of the group represented by the symbol depends upon its position or place value. Thus, 4 is the numeral for a set of four things, but each 4 in the numeral 444 has a different value. Each place has a value 10 times as much as the place on its right:

$$444 = (4 \times 10 \times 10) + (4 \times 10) + (4 \times 1).$$

This could be written as:

$$444 = (4 \times 10^2) + (4 \times 10^1) + (4 \times 1).$$

The 2 and the 1 written just above the 10's are called *exponents*, and show how many 10's are multiplied together.

In the same way, a decimal that expresses a part of a whole thing has place value:

$$.333 = \left(3 \times \frac{1}{10}\right) + \left(3 \times \frac{1}{10} \times \frac{1}{10}\right) + \left(3 \times \frac{1}{10} \times \frac{1}{10} \times \frac{1}{10}\right)$$

Since our system of writing numerals uses ten symbols and the principle of grouping by ten, we say that it is a *base ten* system. It is also called a *decimal* system, the word "decimal" coming from the Latin word for ten.

Our decimal system has two important features. It uses the place value principle that we have just discussed, and it has a symbol for zero. The symbol 0 is used to record "no" things. Zero also serves as a place holder, just like any other digit. Without a zero, it would be impossible to distinguish between numerals like 4 and 40. The features of our numeral system enable us to represent numbers concisely and to perform calculations easily. Simple multiplication and division problems now done by a nine-year-old child in a few minutes took the ablest Roman mathematicians of two thousand years ago several hours to do.

EXERCISE SET 1
Base Ten Numerals

1. What is the value of the 3 in each of these numerals?
   a. 403,157   b. .0143   c. 2.372

2. What is the largest whole number you can write using the digits 2, 3, 4?

3. What is the smallest whole number you can write using the digits 4, 3, 2?

4. Make tally marks and group them to show a picture of place value for 137.

5. Show the value of each digit in these numbers by showing the multiple of ten each represents. For example:

$$478 = (4 \times 10 \times 10) + (7 \times 10) + (8 \times 1)$$

   a. 5,684   b. 2,070   c. .413

6. a. Can you think of a reason why ten is a base for many numeral systems?

   b. The Maya Indians of Central America had a numeral system with a base of twenty. Can you think of a reason for their selection of twenty as a base?

Numeration with Five Symbols

Counting with Fives

Picture a primitive man, spear in hand, stalking a herd of wild animals. When he overtakes the herd, he makes a count of animals, for he wants to report on this herd to his village. Since he has a spear in one hand, he counts on the fingers of the other hand only. Perhaps it was in this way that some primitive peoples developed base five numeral systems.

Let's try to build a base five numeral system. We will need five number symbols, so let's use the numerals 0, 1, 2, 3, 4. When we count in this system, we will group by fives.

$$\boxed{x\ \ x\ \ x\ \ x\ \ x} \qquad x\ \ x\ \ x$$
$$\boxed{x\ \ x\ \ x\ \ x\ \ x}$$

Figure 3

The number of x's in Figure 3 are separated into two groups of five with 3 left over. In the base five system, we would write this 23_{five}. We write the "five" to show that this number is not "twenty-three" as we usually think of it in base ten. It is 2 fives and 3 ones. The numeral in the second place to the left tells us how many groups of five are tallied. In our base ten or decimal numerals, the number of objects would be written as 13, which means one group of ten and three more. In a similar way, 41_{five} means $(4 \times five) + 1$, or 4 groups of five and one more. And, 324_{five} means $(3 \times five \times five) + (2 \times five) + 4$. The value of each digit place in a base five numeral is five times as much as the place on its right. We can invent numerals which will have any place value we may want them to have.

Whenever we write a numeral like 24, we will consider it a base ten numeral. Numerals in other bases will have the base indicated to the right and below the numeral, as shown in the following examples. These indicators are called *subscripts*.

$$24 \quad = (2 \times 10) + 4$$
$$24_{five} = (2 \times five) + 4$$

Let's make a closer comparison of base ten and base five numerals.

Base ten notation	Base five grouping	Base five notation	Base five names
0		0	zero
1	x	1	one
2	xx	2	two
3	xxx	3	three
4	xxxx	4	four
5	(xxxxx)	10_{five}	one five
6	(xxxxx) x	11_{five}	one five and one
7	(xxxxx) xx	12_{five}	one five and two
8	(xxxxx) xxx	13_{five}	one five and three
9	(xxxxx) xxxx	14_{five}	one five and four
10	(xxxxx) (xxxxx)	20_{five}	two-fives
16	(xxxxx) (xxxxx) (xxxxx) x	31_{five}	three-fives and one
24	(xxxxx) (xxxxx) (xxxxx) (xxxxx) xxxx	44_{five}	four-fives and four
25	[(xxxxx) (xxxxx) (xxxxx) (xxxxx) (xxxxx)]	100_{five}	five-squared
31	[(xxxxx) (xxxxx) (xxxxx) (xxxxx) (xxxxx)] (xxxxx) x	111_{five}	five-squared, one five and one
50	[(xxxxx) (xxxxx) (xxxxx) (xxxxx) (xxxxx)] [(xxxxx) (xxxxx) (xxxxx) (xxxxx) (xxxxx)]	200_{five}	two-five-squared

When a base five numeral is written according to the place value of each symbol, it can easily be changed to base ten notation.

$$342_{\text{five}} = (3 \times \text{five} \times \text{five}) + (4 \times \text{five}) + 2$$
$$= (3 \times 25) + (4 \times 5) + 2 \quad \text{[base ten numerals]}$$
$$= 75 + 20 + 2 = 97$$

A good way to keep place value straight when changing from base five to base ten is to refer to quarters, nickels, and pennies. For example, 342_{five} can be interpreted as:

$(3 \times \text{five} \times \text{five}) + (4 \times \text{five}) + 2,$
or 3 quarters + 4 nickels + 2 pennies,
or 75 cents + 20 cents + 2 cents = 97 cents.

Another way to remember the place value of base five numerals is to write the symbols in columns. Let's use this method to get base ten interpretations for 34_{five}, 241_{five}, 303_{five}, $1{,}420_{\text{five}}$, and $3{,}042_{\text{five}}$.

five x five x five (five³)	five x five (five²) or "quarters"	five x one or "nickels"	ones or "pennies"	
		3	4	$34_{\text{five}} = (3 \times \text{five}) + 4$ $= 19$
	2	4	1	$241_{\text{five}} = (2 \times \text{five} \times \text{five}) +$ $(4 \times \text{five}) + 1$ $= 71$
	3	0	3	$303_{\text{five}} = (3 \times \text{five} \times \text{five}) +$ $(0 \times \text{five}) + 3$ $= 78$
1	4	2	0	$1{,}420_{\text{five}} = (1 \times \text{five} \times \text{five} \times$ $\text{five}) + (4 \times \text{five} \times$ $\text{five}) + (2 \times \text{five}) + 0$ $= 235$
3	0	4	2	$3{,}042_{\text{five}} = (3 \times \text{five} \times \text{five} \times$ $\text{five}) + (0 \times \text{five} \times$ $\text{five}) + (4 \times \text{five}) + 2$ $= 397$

Understanding Base Five Numerals

1. Draw x's and encircle groups to show the meaning of each of the following numerals in base ten:
    a. 14   b. 23   c. 137

2. Draw x's and encircle groups to show the meaning of each numeral:
    a. 14_{five} b. 23_{five} c. 134_{five}

3. Write the meaning of each of these base ten numerals according to place value; for example, $44 = (4 \times 10) + 4$:
    a. 53   b. 123   c. 2,304

4. Write the meaning of each of these base five numerals according to place value; for example, $23_{five} = (2 \times five) + 3$:
    a. 43_{five} b. 123_{five} c. $2,304_{five}$

5. Write the meaning of each of these base five numerals and change to base ten notation:
    a. 24_{five} b. 441_{five} c. $1,203_{five}$

Changing from Base Ten to Base Five

We have already changed numbers written in base five to base ten numerals. Let's now see if we can change from base ten to base five. Let's try to write 89 as a base five numeral.

In base five, the values of the places are one, five, five $\times$ five, five $\times$ five $\times$ five, etc. In base ten, these place values are:

$$one = 1$$
$$five = 5$$
$$five \times five = 25$$
$$five \times five \times five = 125$$

Now, 89 is less than 125; hence it is less than a four-place base five number. But it is more than 25, so it can be written as a three-place base five numeral.

$$\begin{array}{r} 3 \\ 25\overline{)89} \\ 75 \\ \hline 14 \end{array}$$ The division shows that there are 3 five $\times$ five groups in 89, and 14 units left over.

$$\begin{array}{r} 2 \\ 5\overline{)14} \\ 10 \\ \hline 4 \end{array}$$ This division shows that there are 2 five groups in 14, and 4 units left over.

We can see that:
$$89 = (3 \times \text{five} \times \text{five}) + (2 \times \text{five}) + 4$$
$$= 324_{\text{five}}$$

We can think of this process as one in which the base ten numeral is broken down into multiples of five. Thus:
$$317 = 250 + 50 + 15 + 2$$
$$= (2 \times \text{five} \times \text{five} \times \text{five}) + (2 \times \text{five} \times \text{five})$$
$$+ (3 \times \text{five}) + 2$$
$$= 2{,}232_{\text{five}}$$

A shortcut for changing base ten to base five is to make continued divisions. (In the following example, division is indicated with this sign: $\underline{)\quad}$.)

5$\overline{)317}$ **Remainders**

5$\underline{)\ 63}$ groups of five – – – – – – – →2 ones left over

5$\underline{)\ 12}$ groups of five × five – – – – →3 fives left over

    2 groups of five × five × five – –→2 groups of five × five
                                                    left over

Writing the remainders in reverse order gives the result $317 = 2{,}232_{\text{five}}$.

EXERCISE SET 3
Changing to and from Base Five

1. Change these base ten numerals to base five:
    a. 38   b. 64   c. 156   d. 377

2. Change these base five numerals to base ten:
    a. 23_{five} b. 222_{five} c. 340_{five} d. $2{,}003_{\text{five}}$

A Base Five Counting Board

In many Oriental shops today, no cash registers or adding machines are seen. Instead, the merchant totals his sales on an instrument consisting of beads that can be made to slide on slender bamboo rods. Such an instrument is pictured in Figure 4. It is called a counting board or *abacus*. One type of abacus was used in China as early as the sixth century B.C., and a form of this counting board has been used at some time in almost every area of the world.

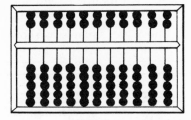

Figure 4

You can make a base five abacus that will illustrate place value in base five and help in computations with base five numerals. You will need these materials: one $1'' \times 1'' \times 8''$ wood stick, 5 long nails, twenty-five $1'' \times 1''$ cardboard squares or beads (5 of each color if possible).

To construct this abacus, first locate five equally-spaced points about $1\frac{1}{2}''$ apart along the middle of one side of the stick, as shown in Figure 5a. Punch holes in the center of each cardboard square. Place five cardboard squares of the same color on each nail and pound the nails into the sticks at the marked points.

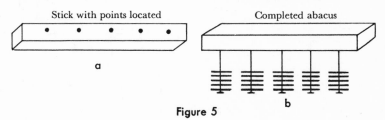

Stick with points located

Completed abacus

a

b

Figure 5

This counting board can be used to represent base five numerals. Figure 6 shows how we can use the abacus to represent 12_{five}.

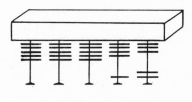

Figure 6

We can now use the abacus for base five computations.

Addition of Numbers Written in Base Five

You know that the sum of 12 and 22 is 34, but what is the sum of 12_{five} and 22_{five}? Since addition is essentially a form of counting, we can use the base five counting board to do this sum. To add 12_{five} and 22_{five}, move the squares as shown in Figure 7.

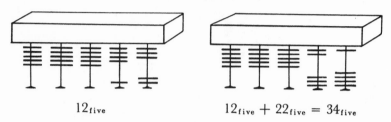

12_{five} $12_{five} + 22_{five} = 34_{five}$

Figure 7

We see that the result is comparable to that obtained from a base ten computation of 12 and 22.

Now try $23_{five} + 14_{five}$. The abacus drawings in Figure 8 show that when you add 3_{five} and 4_{five}, you "carry" a five to the second digit place.

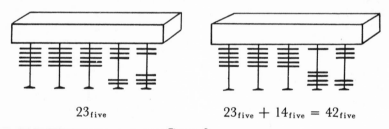

23_{five} $23_{five} + 14_{five} = 42_{five}$

Figure 8

Another way to add small numbers in base five numeration is to count on a number line. The base five number line in Figure 9 shows how to add $4_{five} + 3_{five}$.

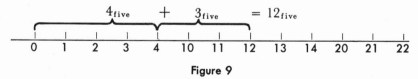

Figure 9

If you want to become proficient in adding numbers written in base five, you should make an addition table for reference or else memorize all the addition combinations, just as you did in school for base ten. A good way to check your work is to translate to base ten and compare your results as shown in the following examples. Notice the similarity of the base five and base ten addition processes.

A.   **Check**

23_{five} =   2 fives and 3 ones =   13 = 1 ten and 3 ones
$+ 13_{five}$ = + 1 five and 3 ones = + 8 = +   8 ones
41_{five} =   3 fives and 6 ones =   21 = 1 ten and 11 ones
                 or   or
                 4 fives and 1 one   2 tens and 1 one

B.   **Check**   **C.**   **Check**

24_{five} =   14 32_{five} =   17
$+ 13_{five}$ = + 8 14_{five} =   9
42_{five} =   22 $+ 43_{five}$ =   + 23
                                144_{five} =   49

EXERCISE SET 4
Base Five Addition

1. Copy and complete this addition table for numbers written in base five:

+	1	2	3	4	10
1	2				
2		4			
3		10	11	12	
4				13	
10					20

2. Use this table to perform these additions in base five.

a. 23_{five} b. 22_{five} c. 43_{five} d. 244_{five}
  $+ 21_{five}$ $+ 14_{five}$ $+ 14_{five}$ $+ 204_{five}$

3. How much do you "carry" when you add these numbers in base ten?   59
                                                27

4. How much do you "carry" when you add these numbers in base five? 23_{five}
                                                14_{five}

Subtraction in Base Five

$12_{five} - 4_{five} = ?_{five}$

What is subtraction? You probably know that subtraction is the inverse of addition. In other words, in base ten if $3 + 4 = 7$, then $7 - 4 = 3$ and $7 - 3 = 4$, or if $a + b = c$, then $c - b = a$ and $c - a = b$. In the subtraction $7 - 4$, we are asking what number added to 4 gives 7.

If we know the addition combinations for numbers written in base five, we can easily subtract numbers in base five. In the subtraction problem $12_{five} - 4_{five}$, we are asking what number added to 4_{five} will give 12_{five}. Looking in the addition table of Exercise Set 4, we find 12_{five} in row 3. This tells us that $4_{five} + 3_{five} = 12_{five}$, and we therefore know that $12_{five} - 4_{five} = 3_{five}$.

Study the following subtraction examples, and check the results by translating to base ten. Notice the similarity between the base five and base ten subtraction methods.

A.

40_{five} = 4 fives and 0 ones =	20 = 2 tens and 0 ones =	1 ten and 10 ones
$- 3_{five}$ = $-$　　　　3 ones =	$- 3 = -$　　　　3 ones =	$-$　　　　3 ones
32_{five} =	17 =	1 ten and　7 ones

B.　　　　**C.**　　　　**D.**　　　　**E.**　　　　　　**Check**

14_{five}	11_{five}	142_{five}	403_{five} =	103
$- 3_{five}$	$- 4_{five}$	$- 24_{five}$	$- 134_{five}$ =	$- 44$
11_{five}	2_{five}	113_{five}	214_{five} =	59

Another way to do base five subtraction is to label a number line with base five numerals and count backward.

$$12_{five} - 4_{five} = 3_{five}$$

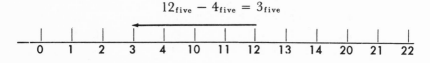

Figure 10

Then $12_{five} - 4_{five}$ becomes a matter of beginning at 12_{five} and counting 4 units to the left. This brings us to the answer, 3_{five}. Or, if we start at 4_{five} and count the units to 12_{five}, we find that the difference between the two is 3_{five}.

1. Perform the following subtractions. Check by adding with the base five abacus or by translating to base ten:

a. 44_{five}	b. 32_{five}	c. 320_{five}	d. $4,103_{\text{five}}$
$-\ 23_{\text{five}}$	$-\ 14_{\text{five}}$	$-\ 42_{\text{five}}$	$-\ 424_{\text{five}}$

2. When we "borrow" in the problem $32 - 14$, with base ten numerals, how many units do we "borrow?"

3. When we "borrow" in base five, in the problem $32_{\text{five}} - 14_{\text{five}}$, how many units do we "borrow?"

Base Five Addition and Subtraction with a Nomograph

If you have trouble doing additions or subtractions involving base five numerals, the chart in Figure 11, called a *nomograph*, can help you.

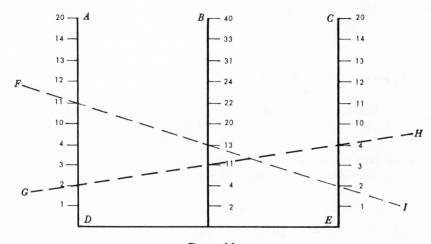

Figure 11

A nomograph is very easy to make. First draw a horizontal line *DE* on a sheet of graph paper. Then draw three parallel, vertical lines the same distance apart to intersect line *DE*. Label the lines *A*, *B*, and *C*, as illustrated. Mark off equal spaces on lines *A*, *B*, and *C*. Starting with *D* and *E* as zero points, number the corresponding marks on lines *A* and *C* consecutively with base five numerals. Number center line B in base five also, but give each mark on this line a numerical value twice that of the corresponding marks on scales A and C.

With a straight edge, connect any point on A with any point on C. The sum of the two numbers represented by the points will be found at the point where the straight edge crosses the center line B. The A reading plus the C reading is equal to the B reading, or $A + C = B$. For example, line FI shows that $11_{five} + 2_{five} = 13_{five}$. The nomograph can also be used for subtraction, for $A = B - C$ and $C = B - A$. Line GH shows that $11_{five} - 4_{five} = 2_{five}$ and $11_{five} - 2_{five} = 4_{five}$. If you have studied plane geometry, you can easily show why the nomograph "works."

EXERCISE SET 6
Working with the Base Five Nomograph

1. Use your nomograph to find the following sums:
   a. $12_{five} + 3_{five}$ b. $11_{five} + 14_{five}$ c. $13_{five} + 13_{five}$
2. Use your nomograph to find the following differences:
   a. $13_{five} - 10_{five}$ b. $11_{five} - 3_{five}$ c. $23_{five} - 14_{five}$

Multiplication in Base Five

If you were given the problem, "Find the sum of $3 + 3 + 3 + 3$ in base ten," you would probably take a short cut and use the multiplication fact $4 \times 3 = 12$. We can make short cuts for repeated additions in base five by constructing a multiplication table with base five numerals. For example, $3_{five} + 3_{five} + 3_{five} + 3_{five} = 4_{five} \times 3_{five} = 22_{five}$. Copy and complete the following base five multiplication table.

Multiplication Table — Base Five

×	1	2	3	4	10
1	1	2			
2	2	4			
3			14		
4	4		31		
10	10				100

Napier's Bones for Base Five Multiplication

In 1617, a Scottish mathematician named John Napier introduced a device, now called Napier's bones, for performing multiplications. We can make a set of Napier's bones to do multiplications in base five.

Cut six strips of cardboard (or plywood), each one inch by five inches. Mark five strips into squares and triangles as shown in Figure 12a, and label these five strips with the base five products for each digit, as shown in Figure 12b. Make the remaining strip an index, as shown in Figure 12c.

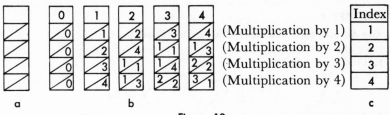

a   b   c

Figure 12

We can use these strips to solve multiplication problems, such as $342_{\text{five}} \times 4$. Select the strips or "bones" for 3, 4, and 2, and place them in order next to the index, as in Figure 13. Since we are multiplying by 4, we obtain the product from the 4 row. Bring down the 3 on the extreme right, and then add the remaining figures diagonally in order to get the rest of the digits in the product.

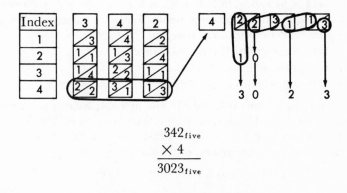

$$342_{\text{five}}$$
$$\times 4$$
$$\overline{3023_{\text{five}}}$$

Figure 13

Notice the similarity between base ten and base five multiplication processes as you study the following problems. Check the work by using Napier's bones or by translating to base ten numerals.

A.	B.	C.	D.	Check
32_{five}	34_{five}	243_{five}	$41_{five} =$	21
$\times 2$	$\times 3$	$\times 42_{five}$	$\times 24_{five} = \times 14$	
114_{five}	212_{five}	1041	314	84
		2132	1320	210
		22411_{five}	$2134_{five} =$	294

EXERCISE SET 7
Base Five Multiplication and Numeration System Principles

1. Find these products:

    a. 32_{five} b. 43_{five} c. 34_{five} d. 103_{five}

        $\times 3$ $\times 4$ $\times 23_{five}$ $\times 42_{five}$

2. In base ten, we know that the product of two numbers is the same, whether the first is multiplied by the second, or the second is multiplied by the first; for example, $6 \times 4 = 4 \times 6$. Multiplication in base ten is said to be *commutative*, for we can commute (interchange) the members of a product and still obtain the same numerical result.

    a. Is base five multiplication commutative?

    b. Is base ten addition commutative?

    c. Is base five addition commutative?

3. Consider this multiplication: $3 \times 5 \times 6$. We could write the product as $(3 \times 5) \times 6 = 15 \times 6 = 90$, or $3 \times (5 \times 6) = 3 \times 30 = 90$. In both cases, we obtain the same result. This is an example of the *associative principle of multiplication* in base ten.

    a. Is base five multiplication associative?

    b. Is base ten addition associative?

    c. Is base five addition associative?

4. What is the product of 8 and 17? You probably haven't memorized an answer for 8×17, so you might think this way: $8 \times 7 = 56$, carry 5, $8 \times 1 = 8$, $8 + 5 = 13$, and the answer is 136. This method of thinking could also be written in the following way:

$$8 \times 17 = 8(10 + 7) = 8 \times 10 + 8 \times 7 = 80 + 56 = 136$$

This illustrates an important principle of base ten numbers which is usually called the *distributive principle of multiplication over addition*. In the above example, we distributed the multiplier 8 to the 10 and 7. Does this distributive principle apply to base five?

Division in Base Five

If we have a multiplication table, we can use it for division as well as for multiplication. This is based on the fact that division is the inverse of multiplication. Division "undoes" multiplication. If $3 \times 4 = 22_{five}$, then $22_{five} \div 4 = 3$, and $22_{five} \div 3 = 4$, or, in any base, if $a = b \times c$, then $a \div b = c$.

As you study the following examples, note that we do our base five divisions the same way as in the decimal system; that is, by comparing, multiplying, and subtracting. (All the numerals are base five, but the base indicator is not written in the body of the work.)

A.
$$
\begin{array}{r}
103_{five} \; r \; 1 \\
4\overline{)423_{five}} \\
4 \\
\hline
23 \\
22 \\
\hline
1
\end{array}
$$

B.
$$
\begin{array}{r}
114_{five} \; r \; 4 \\
23_{five}\overline{)3241_{five}} \\
23 \\
\hline
44 \\
23 \\
\hline
211 \\
202 \\
\hline
4
\end{array}
$$

EXERCISE SET 8
Base Five Division

Perform the indicated divisions:

1. $3\overline{)341_{five}}$

2. $4\overline{)310_{five}}$

3. $21_{five}\overline{)441_{five}}$

4. $32_{five}\overline{)3433_{five}}$

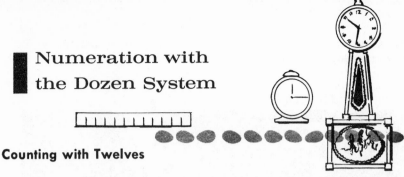

Numeration with the Dozen System

Counting with Twelves

We have seen that it is possible to use a numeration system other than our base ten system. Although we have used five as a base, we could have used any other number. A very important question arises then: "Is there a numeration base better than base ten?" To answer this question, we must consider ways in which a different base could improve upon our present numeration method.

First, it would be practical to select a numeration base having more even divisors than 10 has, for this would simplify work with common fractions. A number like 12 might be a good base, for the even divisors, or factors, of 12 are 2, 3, 4, and 6. The only divisors of 10 are 2 and 5 (besides 10 itself and 1).

It would also be convenient to have a base that is related to our common units of measure. Many of our units of measure are based on 12 or multiples of 12; for example, 12 inches in 1 foot, 12 hours on the clock face, 60 minutes in an hour, 360 degrees in a circle, 12 eggs in a dozen, and 144 units in a gross.

About two hundred years ago, Georges Buffon, a French naturalist, suggested that a base twelve numeration system be universally adopted. Although the base twelve system is sometimes called the "dozen system," we usually call it the *duodecimal system*. *Duodecimal* is another word for twelve, just as *decimal* is another word for ten.

The fight for base twelve was carried into this century, and duodecimal societies sprang up all over the world. Today the Duodecimal Society of America recommends that we use a base twelve system instead of our present decimal notation.

In the duodecimal system we have twelve symbols and form groups of twelves (dozens), twelve-twelves (gross), and so on. This means we must have two new number symbols. The symbols usually used are T for ten and E for eleven.

In the duodecimal system, we group in dozens. This means that the marks in Figure 14 are counted 13_{twelve} because we have 1

Figure 14

group of a dozen and 3 ones. This numeral, 13_{twelve}, is read "one do three," which means one dozen and three.

Let's compare counting in the duodecimal system with counting in the decimal system.

Decimal notation	Duodecimal grouping	Duodecimal notation	Duodecimal names
0		0	zero
1	x	1	one
2	xx	2	two
3	xxx	3	three
4	xxxx	4	four
5	xxxxx	5	five
6	xxxxxx	6	six
7	xxxxxx x	7	seven
8	xxxxxx xx	8	eight
9	xxxxxx xxx	9	nine
10	xxxxxx xxxx	T	ten
11	xxxxxx xxxxx	E	eleven
12	(xxxxxx xxxxxx)	10_{twelve}	one do
13	(xxxxxx xxxxxx) x	11_{twelve}	one do one
23	(xxxxxx xxxxxx) xxxxxx xxxxx	$1E_{\text{twelve}}$	one do eleven
24	(xxxxxx xxxxxx) (xxxxxx xxxxxx)	20_{twelve}	two do
40	(xxxxxx xxxxxx) (xxxxxx xxxxxx) (xxxxxx xxxxxx) xxxx	34_{twelve}	three do four

We can use a place value table to show the relationship between base twelve and base ten numerals. The following table can help us get base ten interpretations for 45_{twelve}, 307_{twelve}, and $12E0_{twelve}$:

twelve x twelve x twelve (twelve³) or dozen-dozen-dozen or great gross (1728)	twelve x twelve (twelve²) or dozen-dozen or gross (144)	twelve x one or dozen or twelves	ones or units
		4	5
	3	0	7
1	2	E	0

$45_{twelve} = (4 \times \text{twelve}) + (5 \times \text{one}) = 53_{ten}$

$307_{twelve} = (3 \times \text{twelve} \times \text{twelve}) + (0 \times \text{twelve}) + 7 \times 1 = 439_{ten}$

$12E0_{twelve} = (1 \times \text{twelve} \times \text{twelve} \times \text{twelve}) + (2 \times \text{twelve} \times \text{twelve}) + (E \times \text{twelve}) + (0 \times \text{one})$

$= (1 \times 1728_{ten}) + (2 \times 144_{ten}) + (11_{ten} \times 12_{ten}) + (0 \times 1)$

$= 2148_{ten}$

EXERCISE SET 9

Working with Base Twelve

Change these numbers, written in base twelve, to base ten.

1. 45_{twelve} **2.** 9_{twelve} **3.** 37_{twelve} **4.** 24_{twelve}

Changing from Base Ten to Base Twelve

If there are advantages in using base twelve numerals, it is important to know how to change numbers written in base ten to base twelve. This conversion is best done by repeated division, in the same way that we changed base ten numerals to base five. Let's change 356 to base twelve.

Check your answers by translating to base ten numerals.

$$
\begin{array}{ll}
\text{A.}\quad 8\,\overline{)54_{\text{twelve}}} \\
\phantom{\text{A.}\quad 8\,)}54
\end{array}
\qquad
\begin{array}{l}
152_{\text{twelve}}\ r\ 3 \\
7\,\overline{)\text{T}05_{\text{twelve}}} \\
7 \\
\overline{30} \\
2\text{E} \\
\overline{15} \\
12 \\
\overline{3}
\end{array}
\qquad
\begin{array}{l}
\phantom{45_{\text{twelve}})}154_{\text{twelve}}\ r\ 40_{\text{twelve}} \\
45_{\text{twelve}}\,\overline{)64\text{T}8_{\text{twelve}}} \\
\phantom{45_{\text{twelve}})}45 \\
\phantom{45_{\text{twelve}})}\overline{1\text{ET}} \\
\phantom{45_{\text{twelve}})}1\text{T}1 \\
\phantom{45_{\text{twelve}})}\overline{198} \\
\phantom{45_{\text{twelve}})}158 \\
\phantom{45_{\text{twelve}})}\overline{40}
\end{array}
$$

Napier's Bones for Duodecimal Numerals

With the base twelve multiplication table, you can now make a set of Napier's bones for performing multiplications in this base. Cut thirteen strips of cardboard or plywood, each one inch by eleven inches. Mark twelve of the strips into twenty triangles and one square. Label each square with a different digit in the base twelve system. Then label the triangles with multiples of the digit in order. Your bone for 8_{twelve} will look like this:

Figure 18

Make one strip an index strip, like this:

Index	2	3	4	5	6	7	8	9	T	E

Figure 19

Index	8	E
2	1 / 4	1 / T
3	2 / 0	2 / 9
4	2 / 8	3 / 8
5	3 / 4	5 / 7
6	4 / 0	5 / 6
7	4 / 8	6 / 5
8	5 / 4	7 / 4
9	6 / 0	8 / 3
T	6 / 8	9 / 2
E	7 / 4	T / 1

Figure 20

Use the bones to multiply in problems such as 8E times T. Place the bones for 8 and E next to the index. Read the product in the T row of the index. Add the numbers along the diagonals to find the product.

EXERCISE SET 13
Base Twelve Multiplication and Division

1. Perform these multiplications with base twelve numerals. Check your answers with Napier's bones.

a. $\begin{array}{r} E \\ \times 7 \\ \hline \end{array}$
c. $\begin{array}{r} 35_{twelve} \\ \times 6 \\ \hline \end{array}$
e. $\begin{array}{r} 4T_{twelve} \\ \times 7 \\ \hline \end{array}$
g. $\begin{array}{r} 46_{twelve} \\ \times T \\ \hline \end{array}$

b. $\begin{array}{r} 23_{twelve} \\ \times 5 \\ \hline \end{array}$
d. $\begin{array}{r} 87_{twelve} \\ \times 9 \\ \hline \end{array}$
f. $\begin{array}{r} 7E_{twelve} \\ \times 8 \\ \hline \end{array}$
h. $\begin{array}{r} TE_{twelve} \\ \times 4 \\ \hline \end{array}$

2. Is T times E = E times T? Does the commutative law for multiplications hold for base twelve numbers?

3. Perform the indicated base twelve divisions. Check by multiplication or translation to base ten.

a. $5\overline{)164_{twelve}}$

c. $T\overline{)542_{twelve}}$

b. $8\overline{)3648_{twelve}}$

d. $23_{twelve}\overline{)57T6_{twelve}}$

Some Fascinating Fractions

Does the symbol $\frac{2}{3}$ have the same meaning in base twelve as in base ten? To answer this, we have to recall what the symbol $\frac{2}{3}$ means. You undoubtedly recognize this symbol as a *common fraction*, an expression of a part of a "whole thing." A drawing like the one in Figure 21 is often used to picture the meaning of $\frac{2}{3}$ It should now be evident that $\frac{2}{3}$ has the same meaning in base twelve as in base ten.

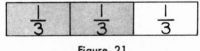

Figure 21

Let us now consider the base twelve fraction $\frac{5}{T}$. If we divide a rectangle into T equal parts and shade 5 of these parts, as in Figure 22, we see that $\frac{5}{T}$ represents one half of the rectangle. If

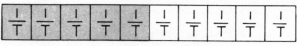

Figure 22

we reduce $\frac{5}{T}$ by division, $\frac{5 \div 5}{T \div 5}$, we find that $\frac{5}{T} = \frac{1}{2}$. This shows how base twelve and base ten fractions may look different but have the same meaning. Fractions in base twelve are different only because duodecimal numerals are used. Similarly,

$$\frac{7}{19_{\text{twelve}}} = \frac{7 \div 7}{19_{\text{twelve}} \div 7} = \frac{1}{3}.$$

By drawings, by reducing to lowest terms, or by knowing the value of duodecimal numerals, you can discover the meaning of a base twelve fraction.

By translating from one number base to another, we can compare fraction values.

Base ten		Base twelve
$\frac{5}{8}$	=	$\frac{5}{8}$
$\frac{7}{10}$	=	$\frac{7}{T}$
$\frac{11}{12}$	=	$\frac{E}{10_{\text{twelve}}}$
$\frac{13}{16}$	=	$\frac{11_{\text{twelve}}}{14_{\text{twelve}}}$

In base ten we find it very handy to use decimal fractions. We change common fractions in base ten numerals to decimals by division. For example, $\frac{3}{4} = 3 \div 4 = .75$. The decimal point is a short way of telling us that the denominator for .75 is one hundred.

In the same way, we change base twelve common fractions to *duodecimal* fractions. Duodecimal fractions are fractions with denominators such as twelve, or one gross. Instead of a decimal point, we use a *duodecimal point* (:). For example, :6 means six twelfths, or $\frac{6}{10_{\text{twelve}}}$; and :$26_{\text{twelve}}$ means $\frac{26_{\text{twelve}}}{100_{\text{twelve}}}$, or $\frac{30}{144}$ in base ten.

Let's express some common fractions as duodecimal fractions. The fraction $\frac{3}{4}$ can be interpreted as 3 divided by 4. The division in base twelve looks like this:

$$4\overline{)3:0_{\text{twelve}}} \quad \begin{array}{l} :9 \quad (\text{Remember } 9 \times 4 = 30_{\text{twelve}}) \\ \underline{3\ 0} \end{array}$$

This tells us that $\frac{3}{4} = :9 = \frac{9}{10_{\text{twelve}}} = \frac{9}{12}$ (in base ten).

Let's use a different method to express $\frac{5}{9}$ as a base twelve rational. We first express $\frac{5}{9}$ as a base ten fraction whose denominator is a multiple of 12.

$$\frac{5}{9} = \frac{80}{144} \quad (\text{in base ten})$$
$$= \frac{68_{\text{twelve}}}{100_{\text{twelve}}}$$
$$= :68$$

One of the reasons base twelve is considered better than our base ten is that, as we have seen, twelve can be divided evenly by 2, 3, 4, and 6. Ten can be divided evenly only by 2 and 5. Thus, there are more base twelve common fractions that can be easily changed to duodecimal fractions than there are base ten common fractions that can be easily changed to decimal fractions.

Try changing other common fractions to duodecimal fractions in the exercises below and you will be surprised at the many simple answers.

EXERCISE SET 14
Working with Base Twelve Fractions

1. Change the following common fractions to duodecimal fractions:

a. $\frac{2}{3}$ b. $\frac{3}{2}$ c. $\frac{5}{6}$ d. $\frac{1}{4}$ e. $\frac{4}{12}$ f. $\frac{3}{8}$

2. Change the following from duodecimal to base ten fractions:

a. :5   b. :6   c. :8   d. :4   e. :24   f. 3:T

3. What denominators for base twelve fractions will give one-digit duodecimal equivalents?

Using Duodecimal Fractions

Duodecimals are very convenient in working with measures based on twelve, such as feet and inches. For example, 5 feet 7 inches = 5:7 feet, or 57_{twelve} inches. This simplifies computing with measured quantities like this:

Base ten addition	Base twelve addition
3 ft. 7 in.	$3:7_{\text{twelve}}$ ft.
+15 ft. 8 in.	+ $13:8_{\text{twelve}}$ ft.
18 ft. 15 in.	$17:3_{\text{twelve}}$ ft.
or 19 ft. 3 in.	

In base ten notation, comparisons involving division are often expressed in per cent. For example, 25% is interpreted as $\dfrac{25}{100}$. In the duodecimal system, *per gross* is used instead of per cent. For example, 25_{twelve} per gross means $\dfrac{25_{\text{twelve}}}{100_{\text{twelve}}}$

If we translate to base ten numerals, $\dfrac{25_{\text{twelve}}}{100_{\text{twelve}}}$ becomes $\dfrac{29_{\text{ten}}}{144_{\text{ten}}}$

Computations are done the same way in both systems. When working with interest rates based on a twelve-month year, the duodecimal system is very convenient. But don't consider 6 per cent and 6 per gross as being equivalent rates! Which of these rates is the higher?

EXERCISE SET 15
Measurements in Base Twelve

1. Change the following measurements to duodecimal notation and compute as indicated. Check by using the usual computational method.

a.   7 ft. 8 in.
   + 8 ft. 9 in.

c.   3 ft. 4 in.
   × 2 ft. 7 in.

b.   5 ft. 3 in.
   − 2 ft. 8 in.

d. 9 ft. ÷ 2 ft. 3 in.

2. Express 37_{twelve} per gross as an indicated division of base twelve numbers and then change to base ten.

3. Which rate is higher:

a. 12 per cent or 12 per gross?   b. 5 per cent or 8 per gross?

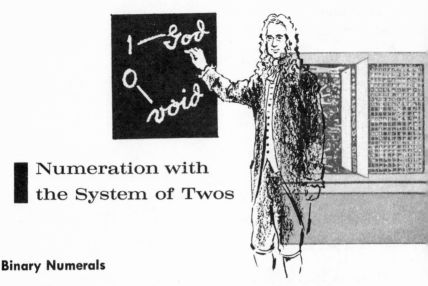

Numeration with the System of Twos

Binary Numerals

During the latter part of the seventeenth century, a great German mathematician, Gottfried Leibniz, advocated the use of the simplest possible numeration system. This system needs only two symbols, 0 and 1, and is called a *binary system*. Leibniz was not the first person to work with a binary system, for certain types of binary systems had been used by even the most primitive peoples. But Leibniz, a very religious man, based his binary numeration system on a religious interpretation. He considered 1 to represent God, and 0 to represent the void, that apart from God. With the two symbols, 0 and 1, Leibniz could represent any number, just as he felt God could fashion the universe from the void.

Binary numerals were almost forgotten after Leibniz's death, but now they have been found to be very useful. The new computing machines, which we sometimes call *electronic brains*, use binary numerals to a great extent. The reason binary numerals are so useful in electronic brains is that 1 can be represented by an electric circuit with the switch "on" and 0 by the switch being "off."

Binary numerals are based on groups of twos, just as our everyday numeration system is based on tens. Let's see how base ten and base two numerals compare.

Base ten notation	Binary grouping	Binary notation	Suggested binary names
1	x	1	one
2		10_{two}	twin
3	x	11_{two}	twin-one
4		100_{two}	one twindred
5	x	101_{two}	one twindred one
6		110_{two}	one twindred twin
7	x	111_{two}	one twindred twin one
8		1000_{two}	one twosand
9	x	1001_{two}	one twosand one
10		1010_{two}	one twosand twin
11	x	1011_{two}	one twosand twin one
12		1100_{two}	one twosand one twindred
13	x	1101_{two}	one twosand one twindred one
14		1110_{two}	one twosand one twindred twin
15	x	1111_{two}	one twosand one twindred twin one
16		10000_{two}	twin twosand

We can use columns to keep place value straight for binary numerals and to change binary numerals to base ten notation. Let's work with 1101_{two}, 10110_{two}, 100000_{two}, and 111111_{two}.

two x two x two x two x two or twindred twosand or thirty-twos	two x two x two x two or twin twosand or sixteens	two x two x two or twosand or eights	two x two or twindred or fours	two x one or twins or twos	ones
		1	1	0	1
	1	0	1	1	0
1	0	0	0	0	0
1	1	1	1	1	1

$1101_{two} = 1(eight) + 1(four) + 0(two) + 1 = 13$

$10110_{two} = 1(sixteen) + 0(eight) + 1(four) + 1(two) + 0$
$= 16 + 4 + 2 = 22$

$100000_{two} = 1(thirty\text{-}two) = 32$

$111111_{two} = 1(thirty\text{-}two) + 1(sixteen) + 1(eight) + 1(four) +$
$1(two) + 1 = 32 + 16 + 8 + 4 + 2 + 1 = 63$

Electric Abacus for Base Two

A binary abacus like the one shown in Figure 23 with two beads or counters will help you keep place value straight and help you

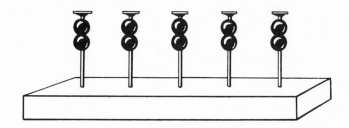

Figure 23

count with binary numerals. An electric abacus made according to the directions below can also be used.

Mount a string of Christmas tree lights evenly spaced on a board, as shown in Figure 24. Be sure the string is the kind where each light will burn independently of the others. When the string is plugged in, bulbs can be turned on or off by screwing the bulbs in or out. When a light is on, it stands for the digit 1. When the light is off, it represents 0.

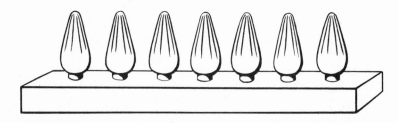

Figure 24

Figure 25 shows how the number 101_{two} looks with this abacus.

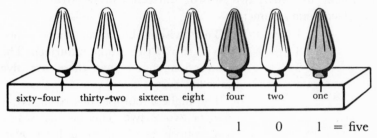

| sixty-four | thirty-two | sixteen | eight | four | two | one |

$$1 \qquad 0 \qquad 1 \ = \ \text{five}$$

Figure 25

EXERCISE SET 16
Working with Binary Numerals

1. Copy and group these marks into twos, to form twins, twindreds, and twosands.

a. xxxxx   b. xxxxx   c. xxxxx   d. xxxx
              xxxxx   xxxxx   xxxxx
                                xxx

2. Write the binary number symbol for each of the numbers represented in Exercise 1 above.

3. Write the binary number names for each of the numbers represented in Exercise 1 above.

4. Make marks and group the marks to show place value for these binary numbers.

a. 110_{two} b. 1011_{two} c. 1100_{two} d. 10101_{two}

5. Write the binary notation for the numbers from 16 to 35.

6. What is the disadvantage in using binary notation?

From Base Ten to Binary Notation

We can change base ten numerals to binary numerals by using methods similar to the ones used to change from base ten to base twelve or base five. We can find the largest multiple of two in the number, subtract this multiple of two, and repeat the process with the remainder. Let's find the binary numeral for 13.

The highest multiple of two in thirteen is eight. Since eight is equivalent to two × two × two, thirteen can be written as a four-place binary numeral.

$$8\overline{)13}$$
$$\quad 1$$
$$\quad 8$$
$$\overline{\quad 5}$$

The division shows that there is one two × two × two group in thirteen. The subtraction shows that there are five units left over.

$$4\overline{)5}$$
$$\quad 1$$
$$\quad 4$$
$$\overline{\quad 1}$$

The highest multiple of two in five is four, or two × two. There is one two × two group in five, and one unit left over. No groups of just two are needed to represent thirteen in binary notation.

We see that $13_{\text{ten}} = 1(\text{two}^3) + 1(\text{two}^2) + 0(\text{two}) + 1 = 1101_{\text{two}}$.

The example below shows how to use a short form of the above method to convert 30_{ten} to binary notation.

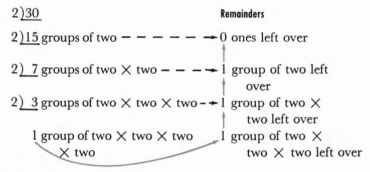

Notice that we performed repeated divisions by 2; the remainders in reverse order gave the binary numeral 11110_{two} as equal to 30_{ten}.

EXERCISE SET 17
Binary Change of Base

1. Change these binary numerals to base ten numerals:
    a. 10101_{two} c. 1011001_{two}
    b. 110011_{two} d. 1111111_{two}

2. Change these base ten numerals to binary notation:
    a. 19   c. 65
    b. 37   d. 100

3. In what digit does an even number always end in binary notation?

Operations in Base Two

After writing the binary notation for numbers such as 30, namely 11110_{two}, you see that binary numerals are very long and tiresome to write. But computation with numbers expressed in the binary system is very simple. In addition, for example, all you need to know is that $1 + 0 = 1$ and $1 + 1 = 10_{two}$. No wonder some persons want us to adopt the binary numeration system! As you study the following examples, check the answers with base ten computation.

A.	B.	C.	D.	E.	Check
1	101_{two}	1	1011_{two}	$1111_{two} =$	15
$+ 1$	$+ 10_{two}$	1	$+ 11_{two}$	$+ 1011_{two} =$	$+ 11$
10_{two}	111_{two}	$+ 1$	1110_{two}	$11010_{two} =$	26
		11_{two}			

Since subtraction is the inverse of addition, we can easily subtract binary numerals. To do subtractions quickly, you need to memorize all the subtraction combinations. This is easily done because there are only three: $1 - 1 = 0$, $1 - 0 = 1$, and $10_{two} - 1 = 1$. Notice the simplicity of the following subtraction examples.

A.	B.	C.	D.	E.	Check
10_{two}	11_{two}	101_{two}	1101_{two}	$11001_{two} =$	25
$- 1$	$- 10_{two}$	$- 11_{two}$	$- 110_{two}$	$- 1011_{two} =$	11
1	1	10_{two}	111_{two}	$1110_{two} =$	14

Multiplication in binary numerals is also very simple. All you need to memorize is that $1 \times 0 = 0 \times 1 = 0$ and $1 \times 1 = 1$. You never get a two-digit numeral in binary multiplication of two one-digit numbers. This means you do not have to "carry" when multiplying numbers expressed in binary notation. Study these multiplication examples to see how base ten multiplication ideas are used in base two. Watch your zeros and your additions!

A.	B.	C.	Check
10_{two}	11_{two}	$1101_{two} =$	13
$\times 10_{two}$	$\times 11_{two}$	$\times 101_{two} =$	$\times 5$
100_{two}	11	1101	15
	11	1101	50
	1001_{two}	$1000001_{two} =$	65

Division with binary numerals is the same as division with decimal numerals. This is simple, too, because the quotient figure obtained is always 1 or 0. Study the examples to see how easy it is to do base two division.

A.

$$
\begin{array}{r}
1\ r\ 1 \\
10_{two}\overline{)11_{two}} \\
10 \\ \hline
1
\end{array}
$$

B.

$$
\begin{array}{r}
110_{two} \\
101_{two}\overline{)11110_{two}} \\
101 \\ \hline
101 \\
101 \\ \hline
\end{array}
$$

C.

$$
\begin{array}{r}
1010_{two}\ r\ 11_{two} \\
110_{two}\overline{)111111_{two}} \\
110 \\ \hline
111 \\
110 \\ \hline
11
\end{array}
$$

EXERCISE SET 18
Binary Operations

Perform the indicated operations using binary numerals. Change to decimal numerals to check.

1. a. 110_{two}
   $+ 11_{two}$

   b. 1011_{two}
   $+ 11_{two}$

   c. 1101_{two}
   $+ 111_{two}$

   d. 10101_{two}
   $+ 1011_{two}$

2. a. 111_{two}
   $- 10_{two}$

   b. 101_{two}
   $- 11_{two}$

   c. 11101_{two}
   $- 1011_{two}$

   d. 1001_{two}
   $- 111_{two}$

3. a. 11_{two}
   $\times 10_{two}$

   b. 101_{two}
   $\times 11_{two}$

   c. 111_{two}
   $\times 101_{two}$

   d. 1001_{two}
   $\times 111_{two}$

4. a. $10_{two}\overline{)10_{two}}$

   b. $10_{two}\overline{)1010_{two}}$

   c. $101_{two}\overline{)11110_{two}}$

   d. $110_{two}\overline{)111011_{two}}$

Base Two and Electronic Brains

We said earlier that the binary numeration system is often used by electronic computers, sometimes called electronic brains. These numerals are used because the two digits, 1 and 0, can be represented by a complete or closed circuit (1) or an open circuit (0). When an electric switch is closed, current flows and a light may go on. When a circuit is closed so that electricity flows, we use a 1 to denote this circuit. When a switch is open so that no current flows, we use a 0 to denote this circuit. The illustrations in Figure 26 show these circuits.

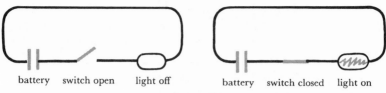

battery switch open light off battery switch closed light on

Figure 26

A number of these circuits arranged in order can represent binary numerals in the same manner as a set of Christmas tree lights in the binary abacus described on page 34.

Electronic brains can add, subtract, multiply, and divide numbers in binary notation as fast as the speed of light. The brains can be given instructions to solve complex mathematical and scientific problems. Computations for a jet engine that would normally require 3,000 hours can be done on an electronic computer in 2 hours. In solving these problems, binary numbers are used as a code to tell the computer what to do.

The actual operation of an electronic brain is very complex, but through a simple example we can get an idea of how the machine uses binary notation. Suppose we want the computer to subtract 13 from 27, and then add 9. Numerical values and instructions are usually put into the computer by magnetic tape or punched cards. We could punch holes in a card to represent

27, 13, and 9, and "feed" the values to the electronic brain. Inside the machine, seven circuits or switches (a, b, c, d, e, f, g) might be arranged in a definite order to represent binary numerals up to seven places. The following table shows how the circuits could represent 27, 13, and 9 in binary notation.

Base ten numerals	Binary numerals	Circuits						
		a	b	c	d	e	f	g
27	0011011			on	on		on	on
13	0001101				on	on		on
9	0001001				on			on

After "feeding" the numbers to the machine, we would have to tell the machine what to do with the numbers. Operational instructions, such as those shown in red below, would have to be given to the computer so that it could work the problem in a logical manner.

1. Subtract two numbers. ⌐⌐⌐ ⌐ **2.** Write the difference.

$$27 - 13 = 14$$
$$14 + 9 = 23$$

3. Add to the difference. ⌐⌐⌐ **4.** Write the answer.

Each instruction must be represented with a numerical code. The instructions can then be punched on a card in the numerical code and fed into the computer in the proper order. Inside the electronic brain, the operational instructions would be changed to a binary code. The table below gives an idea of how the operations might be represented using another set of seven circuits in the computer. The computer is built so that numerals that represent operations can be distinguished from numerals to be operated on.

Operation	Binary numeral code	Circuits						
		a	b	c	d	e	f	g
Subtract two numbers.	0010101			on		on		on
Write the difference.	0001011				on		on	on
Add to the difference.	0110011		on	on			on	on
Write the answer.	1000000	on						

If everything was done properly, the binary numeral 0010111 would be obtained in a computer circuit. This could be translated to 23 and recorded on magnetic tape or punched in a card as the answer.

Electronic brains are no better than the mathematicians who use them. The mathematicians have to work out the problem to be solved in steps so that it can be put into the machine. After the method of solution is worked out, a programmer sets the problem up in code and in logical steps called a *program*. Figure 27 shows the program of a simple problem for one type of computer. In this way, electronic brains are solving all sorts of problems, from finding insurance rates to getting a space ship into orbit.

	STORAGE ENTRY WORDS									
	1	2	3	4	5	6	7	8	9	10
Memory Address	0101	0102	0103	0104						
Input Card	A	B	C	D						

	STORAGE EXIT WORDS									
	1	2	3	4	5	6	7	8	9	10
Memory Addr.	0127	0128	0129	0130	0131					
Output Card	A	B	C	D	T					

Location of Instruction	OP	Data	Instr	Operation Abbrev	8003 Upper Accumulator	8002 Lower Accumulator	8001 Distributor	Remarks
0000	70	0101	0001	RD				READ A CARD
0001	60	0101	0002	RAU	A		A	ENTER A INTO UPPER
0002	24	0127	0003	STD	A		A	STORE A FOR PUNCHING
0003	10	0102	0004	AU	A+B		B	ADD B TO A
0004	24	0128	0005	STD	A+B		B	STORE B FOR PUNCHING
0005	19	0103	0006	MULT		(A+B)×C	C	MULTIPLY (A+B) TIMES C
0006	24	0129	0007	STD		(A+B)×C	C	STORE C FOR PUNCHING
0007	64	0104	0008	DIV RU		T	D	DIVIDE [(A+B)×C] BY D
0008	24	0130	0009	STD		T	D	STORE D FOR PUNCHING
0009	20	0131	0010	STL		T	T	STORE RESULTS (T) FOR PUNCHING
0010	71	0127	0000	PCH				PUNCH A CARD

Figure 27. This is a computing machine program. It is set up to solve the equation
$\frac{(A+B)\times C}{D} = T$. *Reprinted through the courtesy of International Business Machines Incorporated.*

Nim, a Binary Number Game

Base two numerals are very important in the internal operation of a computer, but they also have some other less practical, but just as interesting, applications. One such application is in a game called Nim.

This ancient mathematical game is played by two persons using sticks, matches, toothpicks, beads, coins, or other objects as counters. The counters are placed in piles and the players take turns picking up counters from the piles. The player who picks up the last counter wins. The number of piles used, the number of counters in each pile, and the number of counters picked up at each turn may vary. The players may pick any number of counters at one time from one pile. They must not take counters from more than one pile at a given turn, and at least one counter must be taken on each turn. At the next turn, counters may be selected from a different pile, or, if it still remains, from the same pile. Remember, the player picking up the last counter is the winner. Consider this example:

Place 15 counters in three piles containing 3, 5, and 7 counters, as shown in Figure 28. Base two numerals can be used to plan the strategy.

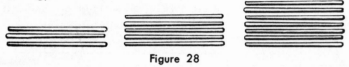

Figure 28

The game is analyzed by writing the number of counters in each pile in binary notation. The winner should always leave the piles so that the sum of the digits of every column of the binary numerals for each pile is an even number. If each sum of the column digits is always even after you pick up a counter, an even number of draws will have to be made to get all the counters, and you will win.

Number of counters in each pile	Binary notation			
	eights	fours	twos	ones
3			1	1
5		1	0	1
7		1	1	1
Column totals (base ten)		2	2	3

In order to win, the columns must be kept even. This may be done by picking up one counter from any pile which will make the column totals 2, 2, 2.

Consider another example. Place 24 counters in three piles, containing 9, 8, and 7 counters, respectively.

Number of counters in each pile		Binary notation
9	=	1001_{two}
8	=	1000_{two}
7	=	111_{two}
Column totals (base ten)		2112

You can put yourself in a winning position by taking six counters (110_{two}) from the third pile. Column totals are then all even, as shown below.

$$9 = 1001_{two}$$
$$8 = 1000_{two}$$
$$1 = 1_{two}$$
$$\overline{2002}$$

EXERCISE SET 19
Questions about Nim

1. If, at the beginning of a game of Nim, column totals are even when you express the counters in binary notation, would you want to draw first or second to insure victory?

2. If the column totals are odd at the start of the game, would you want to draw first or second to insure victory?

More Uses for Base Two

Another interesting application of the binary numeration system is a coin system that simplifies the task of making change. Instead of nickels, dimes, quarters, and half-dollars, a set of coins of value 1, 2, 4, 8, 16, 32, 64 cents is proposed. These new coins would give change for any amount up to $1.28 without using more than one coin of each denomination. For example, 83 cents $= (64 + 16 + 2 + 1)$ cents.

In this new coin system, binary notation would be quite useful in making change. We could write an amount of change in binary notation, and then think of each place in the binary numeral as representing a coin with a value corresponding to the value of the place. A 1 in a place would tell us that a coin of that place value should be used in the change. For example, in binary notation, 83 is 1010011_{two}. This tells us that four coins, a 64, a 16, a 2, and a 1, are to be used to make change for 83 cents. With our present coins, we must often give several coins of the same denomination to make change; for example, 83 cents = 1 half-dollar, 1 quarter, 1 nickel, and 3 pennies. You would need to know binary numbers very well if we had this system of coins.

A clever trick based on binary numbers can be used to guess

your friend's age. For this trick, you need the cards shown in Figure 29.

CARD A		CARD B		CARD C		CARD D	
8	12	4	12	2	10	1	9
9	13	5	13	3	11	3	11
10	14	6	14	6	14	5	13
11	15	7	15	7	15	7	15

Figure 29

Ask your friend to name the cards on which his age is recorded. If he says *A*, *B*, and *D*, you tell him his age is 13. All you have to do to get his age is to add the numbers represented in the upper left-hand corner of the cards he reports. The numerals on the cards he selected are 8, 4, 1, and the sum is 13.

Binary notation is used to make the cards for this trick. Think of each card as representing the value of a place of a four-digit binary numeral. Card *D* represents 1, Card *C* represents two, Card *B* represents four, and Card *A* represents eight. Now write each base ten numeral from 1 to 15 in binary notation. A 1 in a digit place of the binary numeral indicates that the original base ten numeral is to be written on the card that corresponds to the value of that digit place. Thus, $8 = 1000_{two}$, $4 = 100_{two}$, $2 = 10_{two}$, and $1 = 1_{two}$, and 8, 4, 2, and 1 are written in the upper left-hand corner of cards *A*, *B*, *C*, and *D* respectively. Now 13 is written as 1101_{two}, and thus 13 must appear on Card *A*, Card *B*, and Card *D*. When your friend selects cards *A*, *B*, and *D*, you merely compute $8 + 4 + 1 = 13$.

EXERCISE SET 20
Using Base Two Numerals

1. With a set of coins of value 1, 2, 4, 8, 16, 32, and 64, what coins would be needed to make change for:
     a. $.46?   b. $1.02?   c. $1.12?

2. Make a set of cards that can be used to guess ages from 1 to 31.

3. The binary numbers can be used to make up a system of weights for a scale balance. What weights are needed to weigh any amount from 1 to 15 ounces? Be sure to find the fewest possible weights; no duplications are allowed.

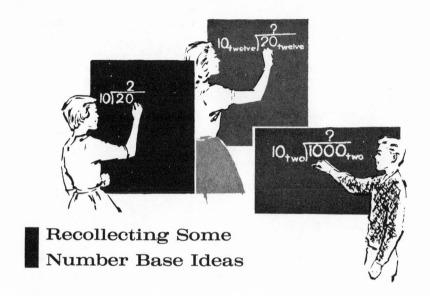

Recollecting Some Number Base Ideas

Division Patterns

How can you tell which numbers written in base ten are divisible by 10, the base of the system? This question is an easy one, for you know that a number written in base ten is divisible by 10 if the units digit of its numeral is zero. Similarly, binary numbers whose numerals end in zero are known to be even numbers or numbers divisible by the base number two. Is it true that any number whose numeral ends in zero is divisible by its base number?

We can find other interesting tests for division in different number bases. Consider these four numbers written in base ten: 117, 414, 963, and 7884. Each number is divisible by 9. Now find the sum of the digits of each numeral and note your result. If you did your work correctly, you found that in each case the sum of the digits was 9 or a multiple of 9. We can show that this is not just a coincidence by examining the numeral 7884. This numeral could be written as

$$(7 \times 1000) + (8 \times 100) + (8 \times 10) + 4$$

or,

$$[7 \times (999 + 1)] + [8 \times (99 + 1)] + [8 \times (9 + 1)] + 4.$$

Applying the distributive principle, we have

$$(7 \times 999) + 7 + (8 \times 99) + 8 + (8 \times 9) + 8 + 4,$$

and, using the associative principle, we can rearrange this to give
$$(7 \times 999 + 8 \times 99 + 8 \times 9) + 7 + 8 + 8 + 4.$$

Now, 999, 99, and 9 are all divisible by 9, so the expression $(7 \times 999 + 8 \times 99 + 8 \times 9)$ is divisible by 9. Therefore, if the original number is divisible by 9, the sum $7 + 8 + 8 + 4$ must be divisible by 9 (and thus a multiple of 9), and it is.

This same type of thinking can be used to develop rules for certain divisions in any base. The following exercises should enable you to establish some interesting tests for divisibility.

EXERCISE SET 21
Tests for Divisibility

1. How can you tell which numbers written in base twelve are divisible by 6?

2. Write several numbers in base five which are divisible by 2. Is there an easy way to tell whether a number written in base five is divisible by 2?

3. How can you tell if a number written in base five is divisible by 4?

4. How can you tell if a number written in base twelve is divisible by 11?

A Backward Glance and a Look to the Future

In our exciting world of today scientists and scholars are discovering and inventing all kinds of new things. We have wonderful new medicines, new machines, new foods, and new homes. At the same time, mathematicians are devising new mathematics which will make possible many new inventions in the future. This section has described some of the number notations which may have many uses in the world of tomorrow.

The study of different numeration systems should have helped you to better understand our base ten system. You have seen the importance of a principle of grouping in different numeration systems, and can better appreciate the importance of grouping in the base ten system. You have seen how the place-value concept makes numerals of any numeration system easy to use and easy to write. Working with place value in other numeration systems should help you understand its meaning and importance in our base ten system.

You know how to add, subtract, multiply, and divide numbers written in base ten, but you have found that you could perform the same operations in base five, base twelve, and base two. Does $1 + 1$ always equal 2? Of course it does. But sometimes we have different ways of writing the answer. We may say $1 + 1 = 10_{two}$. Nevertheless, the basic principles involved in computations are the same in all numeration systems.

You know that $6 + 4 = 4 + 6$, but does $12_{five} + 14_{five} = 14_{five} + 12_{five}$? The affirmative answer to this question merely points out the fact that certain basic principles of mathematics apply to all numeration systems.

You have seen that our base ten system of numeration is not the only one possible. In fact, any number may be used as a base. Perhaps our system is not even the best. You may be the one to invent a different numeration method that will make many new discoveries possible. Other numeration systems already have many uses, but one of the most important uses is to help us better understand the numeration system we use every day — the base ten system.

EXERCISE SET 22

Making Your Own Numeration Systems

1. Make a new numeration system in which:

$$0 = \square$$

$$1 = \emptyset$$

$$2 = \forall$$

$$3 = \triangle$$

Write the numbers from 1 to 25 in this system. Make up a new set of number words.

2. Try these problems:

a. Add:

b. Subtract:

c. Multiply:

d. Divide:

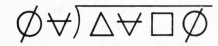

3. Make up new number symbols with a number base other than that illustrated in this booklet.

4. Make up a new numeration system that does not use the principle of place value.

Numeration Review Test

A. Read each of the following statements carefully to determine whether it is a true or false statement:

1. The fraction "$\frac{2}{3}$" has the same value in the dozen system as in base ten.

2. The duodecimal ":5" has the same value as the decimal ".5".

3. We can make a number symbol mean anything that we want it to mean.

4. The numeral "8" (eight) has the same meaning in the duodecimal system as in our decimal system.

5. In performing computations there are more number combinations to remember in the base ten number system than in the base two.

10. When we "carry" a figure in an addition or multiplication problem, the value of the figure we "carry" is the same in base 12 as in base 10.

11. The measurement 3 feet 7 inches can be written 3:7 feet in the duodecimal system.

12. When we borrow in the dozen system in a problem such as 154_{twelve} $- 2T_{\text{twelve}}$, we actually borrow ten units.

13. In the dozen system, 7 times 9 equals 53_{twelve}.

14. In base five, $4 + 3 = 13_{\text{five}}$.

15. The larger the number base the fewer the figures needed to represent large numbers.

20. In the numeral 842_{twelve}, the 8 has a value that is twenty times the value of the 4.

21. In base five, the fraction $\frac{3}{4}$ is equal to $\frac{14_{\text{five}}}{22_{\text{five}}}$.

B. Find the value of the expressions on the left in terms of the base ten numerals on the right. Match these answers in base ten with the numerals on the left. Any choice may be used once, several times, or not at all.

Expressions in base other than ten	Numbers in base ten
_____22. 10101_{two}	A. 31
_____23. 41_{five}	B. 21
_____24. 27_{twelve}	C. 16
_____25. $:6_{\text{twelve}}$	D. .6
_____30. $4_{\text{five}} \times 4_{\text{five}}$	E. .5
_____31. $1001_{\text{two}} + 111_{\text{two}}$	F. .4
_____32. $4_{\text{six}} \div 12_{\text{six}}$	
_____33. $32_{\text{six}} - 4_{\text{six}}$	
_____34. $\dfrac{1}{10_{\text{two}}}$	
_____35. $\dfrac{11_{\text{five}}}{20_{\text{five}}}$	

C. Perform the indicated operations:

40. Write 155_{ten} in base twelve.

41. Write 27_{ten} in base two.

42. Write 434_{ten} in base five.

43. Write 526_{ten} in base six.

44. Write 234_{twelve} in base ten.

45. Write 1324_{five} in base ten.

50. Write 101011_{two} in base ten.

D. Compute as indicated with these base five numerals. Leave the answers in base five.

51. 3202_{five}
$+ 4312_{five}$

53. 431_{five}
$\times 32_{five}$

52. 434_{five}
$- 243_{five}$

54. $4\overline{)121}_{five}$

E. Compute as indicated with these base twelve numerals:

55. $89T7_{twelve}$
$+ E74_{twelve}$

101. 452_{twelve}
$\times T8_{twelve}$

100. $940E_{twelve}$
$- 34T_{twelve}$

102. $8\overline{)57E}_{twelve}$

F. Compute as indicated in base two:

103. 1011_{two}
$+ 110_{two}$

105. 101_{two}
$\times 110_{two}$

104. 11011_{two}
$- 1101_{two}$

110. $10_{two}\overline{)11010}_{two}$

G. Compute as indicated in base six:

111. 342_{six}
$+ 154_{six}$

113. 34_{six}
$\times 5$

112. 530_{six}
$- 42_{six}$

114. $4\overline{)144}_{six}$

H. Perform the following:

115. Change the fraction $\frac{2}{3}$ to a duodecimal.

120. Change the duodecimal :3 to a base ten fraction.

I. Answer the following:

121. In which number base has this addition problem been worked?

$$
\begin{array}{r}
5304 \\
+\ 2334 \\
\hline
12042
\end{array}
$$

122. In which number base has this multiplication been performed?

$$
\begin{array}{r}
123 \\
\times\ 32 \\
\hline
312 \\
1101 \\
\hline
11322
\end{array}
$$

123. What number base has been used in numbering the items of this test?

Extending Your Knowledge

If you have enjoyed exploring the numeration systems, you might want to consult some of these publications for further information:

ADLER, IRVING, *The Magic House of Numbers*. The John Day Co., 1957

BAKST, AARON, *Mathematics, Its Magic and Mastery*. D. Van Nostrand Company, 1952

KASNER, EDWARD and NEWMAN, JAMES, *Mathematics and the Imagination*. Simon and Schuster, 1940

MEYER, JEROME, *Fun with Mathematics*. Dover Publications, 1952

PART II

$$
\begin{array}{r}
1111 \\
\times\,1111 \\
\hline
1111 \\
1111 \\
1111 \\
1111 \\
\hline
1234321
\end{array}
$$

Number
Patterns

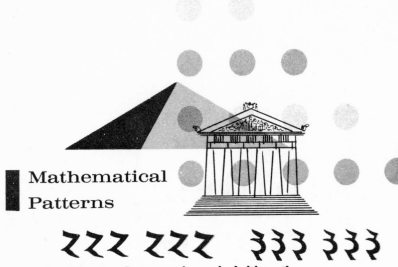

Mathematical Patterns

ZZZ ZZZ ƷƷƷ ƷƷƷ

Discovering Number Patterns through Arithmetic

Through the ages, man has shown an interest in mathematical patterns. The ancient Egyptians and Greeks made extensive use of geometric patterns in their architecture. The Arabs, Hindus, and Greeks worked with number patterns; and some groups even attributed mystical powers to certain number combinations. The mathematicians and scientists of today search for trends or patterns in the numerical results of experiments and problems, for such discoveries often lead to important new ideas. It has even been said that mathematics *is* the study of patterns. One thing is certain — by studying numerical relationships that display unusual patterns, you can learn much about the field of mathematics and have a lot of fun in doing it. The following material can serve as a good starting point.

Look at these relationships:

$$1 \times 9 + 2 = 11$$
$$12 \times 9 + 3 = 111$$
$$123 \times 9 + 4 = 1111$$
$$1234 \times 9 + 5 = 11111$$

If this pattern holds true, you have enough information to write more steps without computing, such as,

$$12345 \times 9 + 6 = 111111$$
$$123456 \times 9 + 7 = 1111111$$

and so on.

The real question is whether or not any of these steps are really true. Of course, you could take each one and follow the directions of multiplying and adding, but that seems like a lot of work. Take a look at just one of these,

$$1234 \times 9 + 5 = 11111,$$

and start to rearrange it.

Write 1234 as

$$1111 + 111 + 11 + 1.$$

Therefore,

$$1234 \times 9 = (1111 + 111 + 11 + 1) \times 9 = 9999 + 999 + 99 + 9.$$

When you add 5 (written as $1 + 1 + 1 + 1 + 1$) to this, you get

$$(9999 + 1) + (999 + 1) + (99 + 1) + (9 + 1) + 1,$$

or

$$10,000 + 1,000 + 100 + 10 + 1,$$

which is equivalent to 11,111.

This is exactly what you wanted to show. This method is general enough, too, to establish that all of the above expressions are true, because any one expression can be handled in exactly the same way. This is the way a mathematician studies problems to save himself a great deal of work.

Such interesting arrangements of numbers would not even exist if you were not able to indicate the relationships by symbols of operation. The number symbols and the operation symbols together form a language — a mathematical language.

EXERCISE SET 1
Working with Number Patterns

On the basis of the number pattern you see in each problem, write at least one more step that follows in the sequence, using the indicated symbols of operation. Notice how these patterns emphasize the principles of grouping by tens and place value that are the basis for our numeral system.

1.
$$1 = 1$$
$$10 + 1 = 11$$
$$100 + 10 + 1 = 111$$
$$1000 + 100 + 10 + 1 = 1111$$
$$\vdots$$

2.
$$10 - 1 = 9$$
$$100 - 1 = 99$$
$$1000 - 1 = 999$$
$$10,000 - 1 = 9999$$
$$\vdots$$

3.
$$(10) = 10$$
$$(10)\,(10) = 100$$
$$(10)\,(10)\,(10) = 1000$$
$$(10)\,(10)\,(10)\,(10) = 10,000$$
$$\vdots$$

4.
$$\frac{1}{(10)} = .1$$
$$\frac{1}{(10)\,(10)} = .01$$
$$\frac{1}{(10)\,(10)\,(10)} = .001$$
$$\frac{1}{(10)\,(10)\,(10)\,(10)} = .0001$$
$$\vdots$$

5.
$$9 + 1 = 10$$
$$90 + 10 = 100$$
$$900 + 100 = 1000$$
$$9000 + 1000 = 10,000$$
$$\vdots$$

6.
$$11(1) = 10 + 1$$
$$11(2) = 20 + 2$$
$$11(3) = 30 + 3$$
$$11(4) = 40 + 4$$
$$\vdots$$

7.
$$1 \times 9 = 10 - 1$$
$$2 \times 9 = 20 - 2$$
$$3 \times 9 = 30 - 3$$
$$4 \times 9 = 40 - 4$$
$$\vdots$$

8.
$$1 \times 8 = 10 - 2$$
$$2 \times 8 = 20 - 4$$
$$3 \times 8 = 30 - 6$$
$$4 \times 8 = 40 - 8$$
$$\vdots$$

9.
$$(1 \times 9) + 2 = 11$$
$$(12 \times 9) + 3 = 111$$
$$(123 \times 9) + 4 = 1111$$
$$(1234 \times 9) + 5 = 11111$$
$$\vdots$$

10.
$$(1 \times 8) + 1 = 9$$
$$(12 \times 8) + 2 = 98$$
$$(123 \times 8) + 3 = 987$$
$$(1234 \times 8) + 4 = 9876$$
$$\vdots$$

11.
$$(1 \times 9) - 1 = 08$$
$$(21 \times 9) - 1 = 188$$
$$(321 \times 9) - 1 = 2888$$
$$(4321 \times 9) - 1 = 38888$$
$$\vdots$$

12.
$$(1 \times 8) - 1 = 07$$
$$(21 \times 8) - 1 = 167$$
$$(321 \times 8) - 1 = 2567$$
$$(4321 \times 8) - 1 = 34567$$
$$\vdots$$

13.
$$9 \times 6 = 54$$
$$99 \times 66 = 6534$$
$$999 \times 666 = 665334$$
$$9999 \times 6666 = 66653334$$
$$\vdots$$

14.
$$7 \times 7 = 49$$
$$67 \times 67 = 4489$$
$$667 \times 667 = 444889$$
$$6667 \times 6667 = 44448889$$
$$\vdots$$

15.
$$6 \times 7 = 42$$
$$66 \times 67 = 4422$$
$$666 \times 667 = 444222$$
$$6666 \times 6667 = 44442222$$
$$\vdots$$

16.
$$333\ 667 \times 111\ 3 = 371\ 371\ 371$$
$$3333\ 6667 \times 111\ 33 = 3711\ 3711\ 3711$$
$$33333\ 66667 \times 111\ 333 = 37111\ 37111\ 37111$$
$$\vdots$$

17.
$$333\ 667 \times 222\ 3 = 741\ 741\ 741$$
$$3333\ 6667 \times 222\ 33 = 7411\ 7411\ 7411$$
$$33333\ 66667 \times 222\ 333 = 74111\ 74111\ 74111$$
$$\vdots$$

18.
$$333\ 667 \times 111\ 6 = 372\ 372\ 372$$
$$3333\ 6667 \times 111\ 66 = 3722\ 3722\ 3722$$
$$33333\ 66667 \times 111\ 666 = 37222\ 37222\ 37222$$
$$\vdots$$

19.
$$333\ 667 \times 222\ 6 = 742\ 742\ 742$$
$$3333\ 6667 \times 222\ 66 = 7422\ 7422\ 7422$$
$$33333\ 66667 \times 222\ 666 = 74222\ 74222\ 74222$$
$$\vdots$$

$$4^2 = (1+1+1+1)^2 = 1 + 2 + 3 + 4 + 3 + 2 + 1$$

Squares of Numbers

You can also build interesting number patterns just by writing numbers in a different way. An unusual example of what can happen when you do this is demonstrated by squaring the whole numbers — but doing this in a very special way.

$$1^2 = 1$$
$$2^2 = (1 + 1)^2 = 1 + 2 + 1$$
$$3^2 = (1 + 1 + 1)^2 = 1 + 2 + 3 + 2 + 1$$
$$4^2 = (1 + 1 + 1 + 1)^2 = 1 + 2 + 3 + 4 + 3 + 2 + 1$$

You can check the correctness of these results by showing one multiplication, that of 4^2, in the following way:

$$4^2 = (1 + 1 + 1 + 1)^2$$

or, in vertical form:

$$
\begin{array}{r}
(1 + 1 + 1 + 1) \\
\times (1 + 1 + 1 + 1) \\
\hline
1 + 1 + 1 + 1 \\
1 + 1 + 1 + 1 \\
1 + 1 + 1 + 1 \\
1 + 1 + 1 + 1 \\
\hline
1 + 2 + 3 + 4 + 3 + 2 + 1
\end{array}
$$

A closely related pattern is the following:

$$1^2 = 1$$
$$11^2 = 121$$
$$111^2 = 12{,}321$$
$$1111^2 = 1{,}234{,}321$$

You can see the similarity to the preceding example by doing the computation of 1111^2 like this:

$$
\begin{array}{r}
1\ 1\ 1\ 1 \\
\times\ 1\ 1\ 1\ 1 \\
\hline
1\ 1\ 1\ 1 \\
1\ 1\ 1\ 1 \\
1\ 1\ 1\ 1 \\
1\ 1\ 1\ 1 \\
\hline
1{,}2\ 3\ 4{,}3\ 2\ 1
\end{array}
$$

Use these patterns to help you fill in the blanks in the next exercise without doing the actual computation.

EXERCISE SET 2

Patterns from the Squares of Integers

Each blank in the following exercises holds a place for one digit in the answer. Fill in the blanks.

1. $\dfrac{22 \times 22}{1 + 2 + 1} = \text{\_ \_ \_}$

2. $\dfrac{333 \times 333}{1 + 2 + 3 + 2 + 1} = \text{\_ \_ \_ \_ \_}$

3. $\dfrac{4444 \times 4444}{1 + 2 + 3 + 4 + 3 + 2 + 1} = \text{\_ \_ \_ \_ \_ \_ \_}$

4. $\dfrac{55555 \times 55555}{1 + 2 + 3 + 4 + 5 + 4 + 3 + 2 + 1} = \text{\_ \_ \_ \_ \_ \_ \_ \_ \_}$

5. $\dfrac{666666 \times 666666}{1 + 2 + 3 + 4 + 5 + 6 + 5 + 4 + 3 + 2 + 1} = \text{\_ \_ \_ \_ \_ \_ \_ \_ \_ \_ \_}$

6. $\dfrac{7777777 \times 7777777}{1 + 2 + 3 + 4 + 5 + 6 + 7 + 6 + 5 + 4 + 3 + 2 + 1} =$

$\text{\_ \_ \_ \_ \_ \_ \_ \_ \_ \_ \_ \_ \_}$

7. $\dfrac{88888888 \times 88888888}{1 + 2 + 3 + 4 + 5 + 6 + 7 + 8 + 7 + 6 + 5 + 4 + 3 + 2 + 1} =$

$\text{\_ \_ \_ \_ \_ \_ \_ \_ \_ \_ \_ \_ \_ \_ \_}$

8. $\dfrac{999999999 \times 999999999}{1 + 2 + 3 + 4 + 5 + 6 + 7 + 8 + 9 + 8 + 7 + 6 + 5 + 4 + 3 + 2 + 1} =$

$\text{\_ \_ \_ \_ \_ \_ \_ \_ \_ \_ \_ \_ \_ \_ \_ \_ \_}$

$(n)(n)-1 = 2(0)$

The Algebra of Number Patterns

A New Method for Discovering Number Relationships

Look now at another number pattern situation. Start with the number 1 and multiply it by itself, obtaining:

$$(1)(1) = 1.$$

Now increase the number 1 by 1 and decrease the number 1 by 1. The product of these two numbers would then be

$$(1 + 1)(1 - 1) = (2)(0) = 0.$$

Repeat this with 2 as the next number to be multiplied by itself.

$$(2)(2) = 4.$$

Increase 2 by 1 and decrease 2 by 1 and multiply:

$$(2 + 1)(2 - 1) = 3(1) = 3.$$

Continue this with 3, 4, 5, and so on, until you see a pattern.

Let's put the answer pairs obtained side by side so the pattern will show up better:

$$\begin{cases} (1)(1) = 1 \\ (2)(0) = 0 \end{cases} \quad \begin{cases} (2)(2) = 4 \\ (3)(1) = 3 \end{cases} \quad \begin{cases} (3)(3) = 9 \\ (4)(2) = 8 \end{cases} \quad \begin{cases} (4)(4) = 16 \\ (5)(3) = 15. \end{cases}$$

Do you notice that the answers on the second line are 1 less than the corresponding answers on the first line? This relationship can be written in the following manner:

$$(1)(1) - 1 = 2(0)$$
$$(2)(2) - 1 = 3(1)$$
$$(3)(3) - 1 = 4(2)$$
$$(4)(4) - 1 = 5(3)$$
$$\vdots$$

A definite pattern is displayed. It should be easy for you to write the next expressions that will follow the same pattern.

It might be advantageous to represent this number pattern without writing a long series of arithmetic expressions that display the pattern. A good way to do this is to let a symbol, such as the letter n, represent any integer. Our number pattern relationship can now be written as

$$(n)(n) - 1 = (n + 1)(n - 1),$$
$$\text{or}$$
$$n^2 - 1 = (n + 1)(n - 1).$$

The symbol n is "holding a place" for any member of the set of positive integers. A symbol that holds a place for any number symbol of a given set of numbers is called a *placeholder* or *variable*.

If we carry out the multiplication indicated in the right member of the above equality statement, we obtain

$$(n + 1)(n - 1) = (n \times n) - n + n - 1 = n^2 - 1.$$

We have shown that each side of our equality statement can be written in the same way, and hence the pattern that it expresses is true for any replacement of n. An algebraic statement that is true for any replacement of the variable involved is called an *identity*.

The process of using variables to represent and establish mathematical relationships is a very important part of the study of algebra.

Another relationship that is very similar to the one presented above is the following pattern:

$$(2)(1) = (1)(1) + 1$$
$$(3)(2) = (2)(2) + 2$$
$$(4)(3) = (3)(3) + 3$$
$$(5)(4) = (4)(4) + 4$$
$$\vdots$$

See if you can write two more steps in the sequence. You can generalize this pattern as follows:

$$(n + 1)n = (n)(n) + n$$
$$\text{or}$$
$$(n + 1)n = n^2 + n.$$

If we perform the multiplication indicated in the left member of this equality statement, we obtain $n^2 + n$, establishing proof for the existence of this number pattern.

Proving Number Patterns

Write one more step for each of the following patterns. Then use the variable n to write a generalization for each pattern. Prove that your generalization is correct.

1.
$$11(1) = 10(1) + 1$$
$$11(2) = 10(2) + 2$$
$$11(3) = 10(3) + 3$$
$$11(4) = 10(4) + 4$$
$$\vdots$$

2.
$$9(1) = 10(1) - 1$$
$$9(2) = 10(2) - 2$$
$$9(3) = 10(3) - 3$$
$$9(4) = 10(4) - 4$$
$$\vdots$$

3.
$$8(1) = 10(1) - 2(1)$$
$$8(2) = 10(2) - 2(2)$$
$$8(3) = 10(3) - 2(3)$$
$$8(4) = 10(4) - 2(4)$$
$$\vdots$$

4.
$$(1 + 1)^2 = 1^2 + 1 + 1 + 1$$
$$(2 + 1)^2 = 2^2 + 2 + 2 + 1$$
$$(3 + 1)^2 = 3^2 + 3 + 3 + 1$$
$$(4 + 1)^2 = 4^2 + 4 + 4 + 1$$
$$\vdots$$

5.
$$1^2 + 1 = 2^2 - 2$$
$$2^2 + 2 = 3^2 - 3$$
$$3^2 + 3 = 4^2 - 4$$
$$4^2 + 4 = 5^2 - 5$$
$$\vdots$$

6.
$$1(5) = \frac{1}{2}(10)$$
$$2(5) = \frac{2}{2}(10)$$
$$3(5) = \frac{3}{2}(10)$$
$$4(5) = \frac{4}{2}(10)$$
$$\vdots$$

7.
$$1(25) = \frac{1}{4}(100)$$
$$2(25) = \frac{2}{4}(100)$$
$$3(25) = \frac{3}{4}(100)$$
$$4(25) = \frac{4}{4}(100)$$
$$\vdots$$

8.
$$15(1) = 1(10) + \frac{1}{2}(10)$$
$$15(2) = 2(10) + \frac{2}{2}(10)$$
$$15(3) = 3(10) + \frac{3}{2}(10)$$
$$15(4) = 4(10) + \frac{4}{2}(10)$$
$$\vdots$$

123456789 ×63

Multiplication problems often provide the basis for many interesting number relationships. Let's examine a few problems of this nature.

Do you have a favorite digit? Let's suppose it is 7. Multiply 7 and 9 to obtain 63. Now multiply:

$$1\ 2,3\ 4\ 5,6\ 7\ 9$$
$$\times 6\ 3$$

Are you surprised at the answer? Try another digit, such as 5. Multiply 5 and 9 to obtain 45. Then multiply:

$$1\ 2,3\ 4\ 5,6\ 7\ 9$$
$$\times 4\ 5$$

Another surprise?

There are only ten digits, 0, 1, 2, 3, 4, 5, 6, 7, 8, 9. You might even try the other eight in the same way. Choose the digit, multiply it by 9, and then multiply this product by

$$1\ 2,3\ 4\ 5,6\ 7\ 9.$$

The answer will have all its digits the same — the one you chose at the beginning.

Can you explain this?

Let's go back to the first example and examine the statement

$$1\ 2,3\ 4\ 5,6\ 7\ 9 \times 63 = 777,777,777.$$

This can be written as

$$1\ 2,3\ 4\ 5,6\ 7\ 9 \times 7 \times 9 = 7(111,111,111).$$

Dividing both sides by 7, we have

$$1\ 2,3\ 4\ 5,6\ 7\ 9 \times 9 = 111,111,111,$$

and it is easy to check that

$$\frac{111,111,111}{9} = 1\ 2,3\ 4\ 5,6\ 7\ 9.$$

If you actually do the division yourself you will see why 8 is missing.

The same reasoning holds for any other digit. Since it is true that

$$1\ 2,3\ 4\ 5,6\ 7\ 9 \times 9 = 111,111,111,$$

then multiplying both sides by the single digit d makes this next statement true:

$$1\ 2,3\ 4\ 5,6\ 7\ 9 \times (d \times 9) = d(111,111,111) = ddd,ddd,ddd.$$

Other special number relations show up with various combinations of products involving 3, 7, 11, 13, and 37. For example,

$$3 \times 37 = 111$$
$$\text{and}$$
$$7 \times 11 \times 13 = 1001$$
$$\text{and}$$
$$3 \times 7 \times 11 \times 13 \times 37 = (111)(1001) = 111,111.$$

EXERCISE SET 4
Multiplication Patterns

On the basis of the few facts given above, see if you can discover an explanation for each of the following number patterns:

1.
$$143 \times 1 = 143 \longrightarrow 143 \times 7 = 1001$$
$$143 \times 2 = 286 \longrightarrow 286 \times 7 = 2002$$
$$143 \times 3 = 429 \longrightarrow 429 \times 7 = 3003$$
$$\vdots$$

Finish the table to
$$143 \times 9 =$$

2.
$$\frac{111}{3} = 37$$

$$\frac{222}{6} = 37$$

$$\frac{333}{9} = 37$$

Write other fractions having the same value with numerators being three 4's, or 5's, or 6's, or 7's, or 8's, or 9's.

3.
$$15873 \times 1 = 15873 \longrightarrow 15873 \times 7 = 111,111$$
$$15873 \times 2 = 31746 \longrightarrow 31746 \times 7 = 222,222$$
$$15873 \times 3 = 47619 \longrightarrow 47619 \times 7 = 333,333$$
$$\vdots$$

Finish the table to
$$15873 \times 9 \times 7$$

Changing Operations

Examine the following number pattern:

$$\left(1\frac{1}{2}\right) \times 3 = 1\frac{1}{2} + 3 = 4\frac{1}{2}$$

$$\left(1\frac{1}{3}\right) \times 4 = 1\frac{1}{3} + 4 = 5\frac{1}{3}$$

$$\left(1\frac{1}{4}\right) \times 5 = 1\frac{1}{4} + 5 = 6\frac{1}{4}$$

$$\left(1\frac{1}{5}\right) \times 6 = 1\frac{1}{5} + 6 = 7\frac{1}{5}$$

Perhaps you have concluded from this that multiplication and addition are identical operations, for you have the same answer whether you multiply or add. Of course, this can't be true in general. But how can you explain this number pattern?

The pattern suggests that the following relationship holds for all values of n:

$$\left(1 + \frac{1}{n}\right) \times (n + 1) = \left(1 + \frac{1}{n}\right) + (n + 1) = (n + 2) + \frac{1}{n}.$$

Notice how this expresses the pattern when you successively let $n = 2$, 3, 4, and 5. Simple algebraic steps show that the three expressions separated by the equal signs are identical.

$$\left(1 + \frac{1}{n}\right) \times (n + 1) = \left(\frac{n}{n} + \frac{1}{n}\right) \times (n + 1)$$

$$= \left(\frac{n + 1}{n}\right) \times (n + 1)$$

$$= \frac{n^2 + 2n + 1}{n}$$

$$\left(1 + \frac{1}{n}\right) + (n + 1) = \frac{n}{n} + \frac{1}{n} + \frac{n^2 + n}{n}$$

$$= \frac{n^2 + 2n + 1}{n}$$

$$(n + 2) + \frac{1}{n} = \frac{n^2 + 2n}{n} + \frac{1}{n}$$

$$= \frac{n^2 + 2n + 1}{n}$$

65

We have demonstrated that the pattern is really expressing a very special relationship. It is saying that you can start with a mixed number of the form $1 + \dfrac{1}{n}$, multiply it by the quantity $n + 1$, or add to it the quantity $n + 1$, and you will get the same result.

EXERCISE SET 5
More Patterns by Changing Operations

The following patterns seem to say that you can perform two different operations on a number and come up with the same answer. From each pattern write the general condition that must be imposed on the numbers to make this possible. Prove that the condition is correct.

1.

Multiply   Subtract

$$1 \times \frac{1}{2} = 1 - \frac{1}{2} = \frac{1}{2}$$

$$2 \times \frac{2}{3} = 2 - \frac{2}{3} = 1\frac{1}{3}$$

$$3 \times \frac{3}{4} = 3 - \frac{3}{4} = 2\frac{1}{4}$$

$$4 \times \frac{4}{5} = 4 - \frac{4}{5} = 3\frac{1}{5}$$

Hint: Show that the following expressions are identical:

$$n\left(\frac{n}{n + 1}\right) = n - \frac{n}{n + 1} = n - 1 + \frac{1}{n + 1}$$

2.

Divide   Add

$$\left(1\frac{1}{3}\right) \div \frac{2}{3} = \left(1\frac{1}{3}\right) + \frac{2}{3} = 2$$

$$\left(2\frac{1}{4}\right) \div \frac{3}{4} = \left(2\frac{1}{4}\right) + \frac{3}{4} = 3$$

$$\left(3\frac{1}{5}\right) \div \frac{4}{5} = \left(3\frac{1}{5}\right) + \frac{4}{5} = 4$$

$$\left(4\frac{1}{6}\right) \div \frac{5}{6} = \left(4\frac{1}{6}\right) + \frac{5}{6} = 5$$

Hint: Let the first expression be:

$$\left(n + \frac{1}{n+2}\right) \div \frac{n+1}{n+2}$$

3.

| Divide | | Subtract | | |

$$\left(4\frac{1}{2}\right) \overset{\text{Divide}}{\div} 3 = \left(4\frac{1}{2}\right) \overset{\text{Subtract}}{-} 3 = 1\frac{1}{2}$$

$$\left(5\frac{1}{3}\right) \div 4 = \left(5\frac{1}{3}\right) - 4 = 1\frac{1}{3}$$

$$\left(6\frac{1}{4}\right) \div 5 = \left(6\frac{1}{4}\right) - 5 = 1\frac{1}{4}$$

$$\left(7\frac{1}{5}\right) \div 6 = \left(7\frac{1}{5}\right) - 6 = 1\frac{1}{5}$$

No hint. This is a do-it-yourself.

From Computations to Number Theories

Experimentation with computations can lead to the discovery of many interesting number relationships. For example, multiply any three consecutive integers,

$$(1)(2)(3),$$
or
$$(2)(3)(4),$$
or
$$(3)(4)(5),$$

and notice that in every case the product is divisible by 6. The reason, of course, is that when three numbers are taken in sequence one of the three must always be an even number and one of them must also be a multiple of 3. The same number might satisfy both conditions. In the product

$$(5)(6)(7),$$

6 is the even number and is also divisible by 3. We can generalize these ideas by saying that

$$(n - 1)(n)(n + 1)$$

is divisible by 6 for all integral values of n.

If we perform the indicated multiplication, we obtain the expression

$$n^3 - n.$$

Therefore, we can say that $(n^3 - n)$ is divisible by 6 for all integral values of n. This tells us that the difference between the cube of a number and the number itself is divisible by 6 for all integral numbers.

Our simple arithmetic computation led us to a general number relationship. A mathematical statement that has been proved is called a *theorem*, and thus we have established a theorem about numbers. There is a special branch of mathematics called the *theory of numbers* which is devoted to the study of number relationships. The theorem we have just established is one of many found in the study of the theory of numbers. It is probable that many number theorems were originally discovered through the recognition of unusual number patterns.

EXERCISE SET 6
Testing Our Number Theorem

Find the values of $(n^3 - n)$ for $n = 2$, $n = 3$, $n = 4$, and $n = 5$, and show that in each case the value found is divisible by 6.

Multiplying Four Consecutive Integers

Let's examine another computational curiosity and attempt to derive a number theorem from it.

Step 1:
Multiply four consecutive integers starting with 1:
$$(1)(2)(3)(4) = 24$$
Add 1: $\qquad 24 + 1 = 25$
Notice that this is a perfect square: $(5 \times 5) = 25$.

Step 2:
Multiply four consecutive integers starting with 2:
$$(2)(3)(4)(5) = 120$$
Add 1: $\qquad 120 + 1 = 121$
Notice that this also is a perfect square: $(11 \times 11) = 121$.

Step 3:

Multiply four consecutive integers starting with 3:
$$(3)(4)(5)(6) = 360$$
Add 1: $\qquad\qquad 360 + 1 = 361$

This also is a perfect square: $(19 \times 19) = 361$.

Have you seen enough to suspect that the product of four consecutive integers plus 1 is *always* a perfect square? It seems too unusual to be possible, but let's see if a proof can be developed.

In general, the product of four consecutive integers starting with n can be written in the form
$$n(n + 1)(n + 2)(n + 3),$$
and this should equal $x^2 - 1$ (1 less than the square of some number x).

Since $x^2 - 1$ can be written as $(x - 1)(x + 1)$, we can complete our proof by showing that $n(n + 1)(n + 2)(n + 3)$ can be written as the product of some quantity minus 1 and the same quantity plus 1.

This is done by regrouping the factors in the following fashion:
$$[n(n + 3)][(n + 1)(n + 2)],$$
or
$$[n^2 + 3n][n^2 + 3n + 2],$$
which can be changed to the form
$$[(n^2 + 3n + 1) - 1][(n^2 + 3n + 1) + 1].$$
This is exactly in the form of
$$(x - 1)(x + 1) \text{ or } x^2 - 1,$$
where
$$x = (n^2 + 3n + 1).$$

EXERCISE SET 7
Testing Our Second Number Theorem

Add 1 to each of the following products and show that they are all perfect squares:

1. $(4)(5)(6)(7)$ $\qquad$ **3.** $(9)(10)(11)(12)$

2. $(5)(6)(7)(8)$

Answers That Show a Pattern

From Any Number to the Same Answer

There is a fascinating set of mathematical problems in which you start with any number, perform a series of arithmetic operations, and always obtain the same answer or answers with similar properties. The following problems are of this nature.

Although it is interesting to note the patterns displayed by the answers obtained in the problems, the real challenge lies in proving the "why" of the problem. Answers and proofs are given for the first five problems, but try to develop the pattern and the proof for each before they are presented.

Although these problems are recreational in nature, they also have a useful aspect, for they stress some important properties of our number system.

Why Is the Answer 22?

Choose any three different numbers less than 10, such as 1, 6, and 8.

Make all the two-digit numbers that you can from these three numbers. You should always get six different numbers. In this case the numbers are:

$$16 \quad 18 \quad 61 \quad 68 \quad 81 \quad 86$$

Add the six numbers:

$$16 + 18 + 61 + 68 + 81 + 86 = 330$$

Add the three original numbers:
$$1 + 6 + 8 = 15$$
Divide the sum of the two-digit numbers by the sum of the one-digit numbers:
$$\frac{330}{15} = 22$$

Now repeat the steps described, using the following sets of three numbers. Look for a pattern in the answers.

a. 1, 2, 3   c. 7, 8, 9
b. 4, 5, 6   d. 2, 4, 6

If you have no errors in computing, you should find that every answer is the same, namely, 22. Let's try to prove that this will always be true.

Let a, b, and c represent any three numbers less than 10. Make all the two-digit numbers that you can from these three numbers. You will always get six numbers which can be represented as

$$10a + b; \ 10a + c; \ 10b + a; \ 10b + c; \ 10c + a; \ 10c + b.$$

Add the six numbers:

$$(10a + b) + (10a + c) + (10b + a) + (10b + c) + (10c + a) + (10c + b)$$
$$= 22a + 22b + 22c$$
$$= 22(a + b + c)$$

When you divide this by the sum of the three numbers, $(a + b + c)$, you obtain 22. The answer is always 22.

Division by 9

Here is a very famous problem that is sometimes used in the arithmetic class.

Select any number, such as 582.
Add the digits: $\quad\quad\quad\quad\quad\quad\quad 5 + 8 + 2 = 15$
Now add the digits of this answer: $\quad 1 + 5 = 6$
Divide the original number by 9:

$$
\begin{array}{r}
6\,4 \\
9\,\overline{)\,5\,8\,2} \\
5\,4 \\
\hline
4\,2 \\
3\,6 \\
\hline
6
\end{array}
$$

Now repeat these steps, using the numbers given below. Continue to add the digits of your answers until the sum is a one-digit number. See if you can discover a pattern in the results.

a. 239   b. 1,053   c. 82   d. 6,975

If you made no errors in computation, you should see a very interesting pattern. If your one-digit sum is less than 9, it is equal to the remainder obtained when dividing the original number by 9. If your one-digit sum is 9, the original number is divisible by 9. This problem is sometimes given the name "casting out nines."

We can use algebra to explain this problem and show that it will always work.

Start with any number (we will select one with four digits):

$$1,000a + 100b + 10c + d.$$

Rewrite the number in this form:

$$a(999 + 1) + b(99 + 1) + c(9 + 1) + d.$$

Group terms as shown:

$$(999a + 99b + 9c) + (a + b + c + d).$$

The expression $999a + 99b + 9c$ contains 9 as a factor and hence is divisible by 9. If the sum of the digits, $a + b + c + d$, is 9, the entire original number contains 9 as a factor and is divisible by 9. If $a + b + c + d$ is less than 9, it is the remainder obtained when the original number is divided by 9. If the sum of the four digits is greater than 9, we can represent the sum as a number in the form $10x + y$. We can then repeat the process described above, obtaining $x(9 + 1) + y = 9x + x + y$.

Now if $x + y$ is 9, the original number is divisible by 9, and if $x + y$ is less than 9, it is the remainder obtained when the original number is divided by 9. If the original number contained more than six digits, more additions might have been necessary to obtain a one-digit number.

The Sum is 3

Here is a problem that is closely related to the previous one. Start with any number, such as 32.

Multiply it by 3:	$3(32) = 96$
Add 1:	$96 + 1 = 97$
Add 1 again:	$97 + 1 = 98$
Add the three numbers:	$96 + 97 + 98 = 291$
Add the digits in the answer:	$2 + 9 + 1 = 12$
Add the digits again:	$1 + 2 = 3$

Repeat these same steps with the following numbers. Keep adding the digits until the sum is just a one-digit number. Do the answers seem to show a pattern?

a. 2   c. 6
b. 4   d. 25

You should have found that all answers were 3. You can easily prove that this is always true.

The last two steps in this problem are identical to the steps in the "division by 9" problem. Thus the preceding problem explains the last two steps of this problem, for taking the sum of the digits of a number will give us the remainder when that number is divided by 9. We will now have to demonstrate that the sum of our three starting numbers, when divided by 9, has a remainder of 3. This can be done as follows:

Start with any number:	N
Multiply it by 3:	$3N$
Add 1:	$3N + 1$
Add 1 again:	$3N + 2$
Add the three numbers:	$3N + (3N + 1) + (3N + 2)$
	$= 9N + 3$

This sum is expressed as a multiple of 9 plus a remainder of 3. When we add the digits of such a number, we would eventually obtain the number 3.

The Three-Digit Reverse and Subtraction

Start with a three-digit number, no two digits the same, such as 532.

Reverse the digits: 235.

Then subtract the smaller from the larger: $532 - 235 = 297$.

Repeat the steps described with the following three-digit numbers and see if you can discover any pattern in the answers.

a. 123   c. 489
b. 956   d. 246

Do even more if you need to.

Have you noticed that in every answer the middle digit is 9, and the other two digits will always add up to 9?

Here is the algebraic proof for these results.

Start with a three-digit number, no two digits the same. Let's use a to represent the hundreds digit, b the tens digit, and c the units digit. Our starting number can then be represented as

$$100a + 10b + c.$$

If we reverse the digits, the new number can be represented as

$$100c + 10b + a.$$

Assuming a is larger than c, we now subtract the smaller from the larger.

$$(100a + 10b + c) - (100c + 10b + a) = 100a - 100c + c - a$$
$$= 100(a - c) + c - a$$
$$= 100(a - c) - (a - c)$$

It looks as if $(a - c)$ represents our new hundreds digit and $- (a - c)$ represents our new units digit. But $- (a - c)$ represents a negative number (we assumed a is greater than c), and it is not satisfactory to write a positive number by using a negative units digit. We can alter this situation by adding 100 and subtracting 100 in the expression obtained above.

$$100(a - c) - (a - c) = 100(a - c) - 100 + 100 - (a - c)$$

This can be rearranged to have a hundreds digit, a tens digit, and a units digit arranged as follows:

$$100(a - c - 1) + 9(10) + [10 - (a - c)].$$

We now have our result in the desirable form. We can see that the tens digit is 9, and the sum of the hundreds and units digits is always 9:

$$(a - c - 1) + [10 - (a - c)] = 10 - 1 = 9.$$

The 1089 Problem

Take any three-digit number, with the hundreds digit differing from the units digit by 2 or more, such as 825.

Reverse the figures: 528

Subtract the smaller from the larger: $825 - 528 = 297$

Reverse the figures of the answer: 792

Add: $792 + 297 = 1089$

Now repeat the steps described, using the following three-digit numbers. Check to see if the answers show a pattern.

a. 123   b. 956   c. 489   d. 246

All of the answers are 1089. Prove that this is so, with all three-digit numbers. Here is a good way to prove it.

Take any three-digit number: $100a + 10b + c$.

Assume: $a > c$ (a is greater than c).

Reverse the figures: $100c + 10b + a$.

Subtract the smaller from the larger: $(100a + 10b + c) - (100c + 10b + a) = 100(a - c) - (a - c)$.

This number must first be expressed as a three-digit number. This is accomplished by using our trick of adding 100 and subtracting 100 and rearranging in the desired form. Thus,

$$100(a-c)-(a-c)$$

can be changed to

$$100(a - c) - 100 + 100 - (a - c)$$

and then to

$$100(a - c - 1) + 9(10) + [10 - (a - c)]. \qquad (1)$$

This is expressed in the form of a three-digit number.

Now reverse the digits to obtain the number:

$$100[10 - (a - c)] + 10(9) + (a - c - 1). \qquad (2)$$

Add these two expressions, (1) and (2):

$100(a - c - 1) + 10(9) + [10 - (a - c)] + 100[10 - (a - c)] + 10(9) + (a - c - 1)$

$= 100[(a - c - 1) + (10 - a + c)] + 10(18) + (10 - a + c) + (a - c - 1)$

$= 100(9) + 10(18) + 9$

$= 100(9) + 10(10 + 8) + 9$

$= 100(9) + 100(1) + 8(10) + 9$

$= 1(1000) + 0(100) + 8(10) + 9$

$= 1089$

It is interesting to notice how the proof shows a special case arising. This occurs when a is just 1 more than c. When this happens, $a - c - 1 = 0$ and $100(a - c) - (a - c)$ always equals 99, which is a two-digit number. You can make this a three-digit number, however, by writing 99 as 099. When this is reversed, you get 990, and the sum, $099 + 990$, is still 1089.

EXERCISE SET 8
More Pattern Problems

Here are two more problems similar to the preceding ones, but the proofs have been omitted.

1. Start with any number, such as   32

   Multiply it by 3: $3(32) = 96$

   Subtract 1: $96 - 1 = 95$

   Subtract 1: $95 - 1 = 94$

   Add the three numbers: $96 + 95 + 94 = 285$

   Add the digits in the answer: $2 + 8 + 5 = 15$

   Add the digits again: $1 + 5 = 6$

Keep adding the digits until the sum is a one-digit number. In this case, you have reached this point when the sum is 6.

Now try this with more numbers such as:

       a. 2   c. 6

       b. 4   d. 25

Correct computing should show that all answers are 6. Prove that this should always be so.

2. Take any two-digit number, such as 43.

   Reverse the digits: 34

   Add the number and its reverse: $43 + 34 = 77$.

Repeat the steps with the following numbers and look for a pattern in the answers.

       a. 12   c. 78

       b. 45   d. 24

Do you notice that in every case the sum is divisible by 11? Prove that this will be true no matter what two-digit number you picked at first.

$$1+3+5+7+9=25$$

$$1=1$$

$$1+3=4$$

$$1+3+5=9$$

$$1+3+5+7+9=25$$

Patterns from Continuous Sums

The Odd Integers Series

Suppose you started to add the consecutive odd numbers:

$$1 + 3 + 5 + \cdots$$

Add one term: $1 = 1$

Add two terms: $1 + 3 = 4$

Add three terms: $1 + 3 + 5 = 9$

Add four terms: $1 + 3 + 5 + 7 = 16$

Add five terms: $1 + 3 + 5 + 7 + 9 = 25$
⋮

A continuous sequence of additions such as this is an example of an *infinite series*. Do you suspect that there is a relationship between the number of terms being added and the sum of these terms?

In the five cases shown, the sum seems to be equal to the number of terms being added multiplied by itself. We can say this in mathematical language in the following fashion:

$$\overbrace{1 + 3 + 5 + \cdots + (\quad)}^{n \text{ terms}} = n^2$$

The question now becomes, "How do you write the nth term?" Perhaps you can discover the pattern here. When you added five terms you had

$$\overbrace{1 + 3 + 5 + 7 + 9}^{\text{five terms}}$$

Notice that $9 = (2 \times 5) - 1$. You might suspect that, in general, when you add n terms your last term to be added is $(2n - 1)$.

It seems reasonable to write

$$1 + 3 + 5 + \cdot \cdot \cdot + (2n - 1) = n^2.$$

The formula works for $n = 1, 2, 3, 4,$ and 5, as you have seen above, and it seems reasonable that it should continue to be true for *all* n, but how can you be sure?

You might feel more sure that this is true if you use geometry to picture what is happening.

Start with one term: Represent this ■
1 by 1 block.

Add two terms: 3 blocks are ■ ■   Notice the new
1 + 3 added to the ■   square array:
previous array. 2 × 2 = 4

Add three terms: 5 blocks are ■ ■ ■   Notice the new
1 + 3 + 5 added to the ■ ■ ■   square array:
previous array. ■ ■ ■   3 × 3 = 9

Add four terms: 7 blocks are ■ ■ ■ ■   Notice the new
1 + 3 + 5 + 7 added to the ■ ■ ■ ■   square array:
previous array. ■ ■ ■ ■   4 × 4 = 16
■ ■ ■ ■

In general, add n terms: $1 + 3 + 5 + \cdot \cdot \cdot + (2n - 1)$

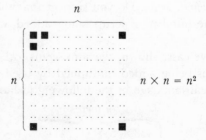

$n \times n = n^2$

Add $(n + 1)$ terms: $(2n + 1)$ blocks are
        added to the previous
        array of n^2 blocks

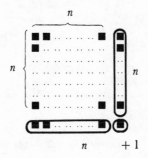

The drawing shows the grand total to be $n^2 + n + n + 1$, or $n^2 + 2n + 1$, which checks out to be exactly what you hoped it would be, namely, $(n + 1)^2$.

There are other ways to look at this sum, too. You might start to add a few consecutive odd integers and begin to notice a way to short-cut the process. For example, in adding the first 10 consecutive odd numbers,

$$1 + 3 + 5 + 7 + 9 + 11 + 13 + 15 + 17 + 19,$$

you might notice how these numbers can be paired to give five sums of 20 each.

$(1 + 19) = 20$ (the sum of the 1st and last)

$(3 + 17) = 20$ (the sum of the 2nd and next to the last)

$(5 + 15) = 20$ (the sum of the 3rd and the 3rd from the last)

$(7 + 13) = 20$ (the sum of the 4th and the 4th from the last)

$(9 + 11) = 20$ (the sum of the 5th and the 5th from the last)

The total sum of the 10 numbers is (5×20), or 100, and this agrees with the answer found by the method of squaring the number of terms being added: $10^2 = 100$.

You could also look at it this way:
    Write the series down twice, with the second row just the reverse of the first row.

$$1 + \ 3 + \ 5 + \ 7 + \ 9 + 11 + 13 + 15 + 17 + 19$$
$$19 + 17 + 15 + 13 + 11 + \ 9 + \ 7 + \ 5 + \ 3 + \ 1$$
The sum is: $\overline{20 + 20 + 20 + 20 + 20 + 20 + 20 + 20 + 20 + 20,}$

which totals (10×20) or 200. But this is twice the desired sum, and so the desired result is half of 200, or 100, again the correct answer.

Still another way to think of this is to see it as a sum of numbers that are equally spaced. The average size could be thought of, then, as the average of the first and the last, that is,

$$\frac{1 + 19}{2} = \frac{20}{2} = 10.$$

Then the sum of 10 numbers, each the size of 10, would be (10×10) or 100. This again checks with all of the other methods.

It would be worth your while to see if you can generalize each of these methods to find that the sum of n terms can be written as

$$1 + 3 + 5 + 7 + \cdot \cdot \cdot + (2n - 1) = n^2.$$

Another interesting feature about this series is that it can be used with computing machines to find the square root of a number. All that you need to do to find the square root of a number is to subtract the consecutive odd numbers and to keep track of the number of subtractions that can be made. This number of subtractions is the square root of the number.

For example, to find the value of $\sqrt{25}$, start subtracting the consecutive odd numbers:

$$\left.\begin{array}{r} 25 - 1 = 24 \\ 24 - 3 = 21 \\ 21 - 5 = 16 \\ 16 - 7 = 9 \\ 9 - 9 = 0 \end{array}\right\} \text{ 5 subtractions}$$

It took 5 subtractions to get from 25 to 0. Therefore,

$$\sqrt{25} = 5.$$

Check this method by taking any perfect square like 100 or 225.

The consecutive odd numbers, strangely enough, are also closely connected to the cubes of whole numbers as well as the squares. The number pattern that shows this is

$$1^3 = 1$$
$$2^3 = 8 = 3 + 5$$
$$3^3 = 27 = 7 + 9 + 11$$
$$4^3 = 64 = 13 + 15 + 17 + 19$$
$$\vdots$$

The cubes are also related to squares of numbers as shown in the following pattern:

$$1^3 = 1^2$$
$$2^3 = (1 + 2)^2 - 1^2$$
$$3^3 = (1 + 2 + 3)^2 - (1 + 2)^2$$
$$4^3 = (1 + 2 + 3 + 4)^2 - (1 + 2 + 3)^2$$
$$\vdots$$

Now note the similarity between this pattern and the one that follows:

$$1^2 = 1$$
$$2^2 = (1 + 2) + 1$$
$$3^2 = (1 + 2 + 3) + (1 + 2)$$
$$4^2 = (1 + 2 + 3 + 4) + (1 + 2 + 3)$$
$$\vdots$$

There seems to be no end to these unusual relationships. A few more are given in the next exercise.

EXERCISE SET 9
Continuous Sums

Find the sums indicated in parts *a.–d.* of the following. Verify that the sums agree with the formula shown in *e.*, when $n = 1, 2, 3,$ and 4.

1. a. $1 =$
   b. $1 + 2 =$
   c. $1 + 2 + 3 =$
   d. $1 + 2 + 3 + 4 =$
   e. $1 + 2 + 3 + 4 + \cdots + n = \dfrac{n(n + 1)}{2}$

2. a. $2 =$
   b. $2 + 4 =$
   c. $2 + 4 + 6 =$
   d. $2 + 4 + 6 + 8 =$
   e. $2 + 4 + 6 + 8 + \cdots + 2n = n(n + 1)$

3. a. $\qquad\quad 1 \qquad\qquad =$
   b. $\qquad 1 + 2 + 1 \qquad\quad =$
   c. $\quad 1 + 2 + 3 + 2 + 1 \qquad =$
   d. $1 + 2 + 3 + 4 + 3 + 2 + 1 \quad =$
   e. $1 + 2 + 3 + 4 + \cdots + n + \cdots + 3 + 2 + 1 = n^2$

4. a. $1^2 =$

   b. $1^2 + 2^2 =$

   c. $1^2 + 2^2 + 3^2 =$

   d. $1^2 + 2^2 + 3^2 + 4^2 =$

   e. $1^2 + 2^2 + 3^2 + 4^2 + \cdots + n^2 = \dfrac{n^3}{3} + \dfrac{n^2}{2} + \dfrac{n}{6}$

$$\text{or} = \frac{n(n+1)(2n+1)}{6}$$

5. a. $1^3 =$

   b. $1^3 + 2^3 =$

   c. $1^3 + 2^3 + 3^3 =$

   d. $1^3 + 2^3 + 3^3 + 4^3 =$

   e. $1^3 + 2^3 + 3^3 + 4^3 + \cdots + n^3 = \dfrac{n^4}{4} + \dfrac{n^3}{2} + \dfrac{n^2}{4}$

$$\text{or} = \frac{n^2(n+1)^2}{4}$$

6. a. $1^3 =$

   b. $1^3 + 3^3 =$

   c. $1^3 + 3^3 + 5^3 =$

   d. $1^3 + 3^3 + 5^3 + 7^3 =$

   e. $1^3 + 3^3 + \cdots + (2n-1)^3 = n^2(2n^2 - 1)$

7. a. $1(2) =$

   b. $1(2) + 2(3) =$

   c. $1(2) + 2(3) + 3(4) =$

   d. $1(2) + 2(3) + 3(4) + 4(5) =$

   e. $1(2) + 2(3) + \cdots + n(n+1) = \dfrac{n}{3}(n+1)(n+2)$

8. a. $\dfrac{1}{1(2)} =$

   b. $\dfrac{1}{1(2)} + \dfrac{1}{2(3)} =$

   c. $\dfrac{1}{1(2)} + \dfrac{1}{2(3)} + \dfrac{1}{3(4)} =$

   d. $\dfrac{1}{1(2)} + \dfrac{1}{2(3)} + \dfrac{1}{3(4)} + \dfrac{1}{4(5)} =$

   e. $\dfrac{1}{1(2)} + \dfrac{1}{2(3)} + \cdots + \dfrac{1}{n(n+1)} = \dfrac{n}{n+1}$

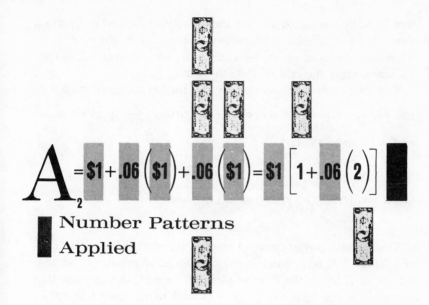

$$A_2 = \$1 + .06 \left(\$1\right) + .06 \left(\$1\right) = \$1 \left[1 + .06 \left(2\right)\right]$$

Number Patterns Applied

How Money Earns Money

Series of numbers like those in the last section are interesting because it seems unusual to add more and more in a series and still have a control on the sum of n terms by means of a formula. The more you think about it, though, you begin to realize that this shouldn't be too much to expect because there is a pattern based on how much you add at each successive stage. Therefore, the sum should also show some kind of pattern, even though it might be too complex to recognize immediately.

Many of the scientific principles that we know today can be traced back to a stage where data indicated a pattern that could be expressed mathematically.

Important financial concepts such as simple and compound interest also show a mathematical pattern. If you invest \$1 at 6% simple interest (only the original \$1 earns interest), the amount the \$1 is worth in successive years could be computed in this fashion:

$$A_1 = \$1 + .06(\$1) = \$1(1 + .06)$$
$$A_2 = \$1 + .06(\$1) + .06(\$1) = \$1[1 + .06(2)]$$
$$A_3 = \$1 + .06(\$1) + .06(\$1) + .06(\$1) = \$1[1 + .06(3)]$$
$$\vdots$$
$$A_n = \$1 + \$1(.06)n = \$1(1 + .06n)$$

The subscripts are used to show the amount at the end of a certain year. Thus A_1 means the amount at the end of the first year; A_2 means the amount at the end of the second year; A_n means the amount at the end of the nth year.

You can generalize the pattern still further for any amount of money, m, invested at a rate of $r\%$, by putting $\frac{r}{100}$ in place of .06 and m in place of \$1:

$$A_n = m\left(1 + \frac{r}{100}n\right).$$

Or, if you let $i = \frac{r}{100}$, the formula becomes

$$A_n = m(1 + i \times n).$$

Many banks pay compound interest on investments. If you invest \$1 at 6% compound interest (interest is reinvested so that interest is paid on the interest already earned), the amount that the \$1 is worth in successive years could be computed like this:

$$A_1 = \$1 + .06(\$1) = \$1(1 + .06)$$
$$A_2 = A_1 + .06\,A_1 = A_1(1 + .06) = \$1(1 + .06)^2$$
$$A_3 = A_2 + .06\,A_2 = A_2(1 + .06) = \$1(1 + .06)^3$$
$$\vdots$$
$$A_n = \$1(1 + .06)^n$$

For m dollars at $r\%$, this generalizes to

$$A_n = m\left(1 + \frac{r}{100}\right)^n.$$

Or, if $i = \frac{r}{100}$, then

$$A_n = m(1 + i)^n.$$

You can see how different simple interest is from compound interest just by looking at the patterns. The number of years, n, is used in entirely different ways.

For those working in banking and investments, there are tables already worked out for different rates of interest for the amount earned by \$1. For any other amount, such as \$500, you first find the amount in the table that \$1 earns and then multiply the value by 500 to find what \$500 earns. The following table shows some selected values for different interest rates for simple as well as compound interest. You can see by number patterns how \$1 increases in value if it is invested under different conditions.

Comparing the Value of $1 Invested at
Simple or Compound Interest for *n* Years

Years	2% (i = .02)		4% (i = .04)		6% (i = .06)	
	Simple	Compound	Simple	Compound	Simple	Compound
1	1.02	1.0200	1.04	1.0400	1.06	1.0600
2	1.04	1.0404	1.08	1.0816	1.12	1.1236
3	1.06	1.0612	1.12	1.1249	1.18	1.1910
4	1.08	1.0824	1.16	1.1699	1.24	1.2625
5	1.10	1.1041	1.20	1.2167	1.30	1.3382
• • •						
10	1.20	1.2190	1.40	1.4802	1.60	1.7908
• • •						
16	1.32	1.3728	1.64	1.8730	1.96	2.5404
17	1.34	1.4002	1.68	1.9479	2.02	2.6928
18	1.36	1.4282	1.72	2.0258	2.08	2.8543
19	1.38	1.4568	1.76	2.1068	2.14	3.0256
20	1.40	1.4859	1.80	2.1911	2.20	3.2071
• • •						
30	1.60	1.8114	2.20	3.2434	2.80	5.7435
• • •						
40	1.80	2.2080	2.60	4.8010	3.40	10.2857
• • •						
50	2.00	2.6916	3.00	7.1067	4.00	18.4202
• • •						
n	$(1 + .02n)$	$(1 + .02)^n$	$(1 + .04n)$	$(1 + .04)^n$	$(1 + .06n)$	$(1 + .06)^n$

EXERCISE SET 10
Simple and Compound Interest

Use the interest table to answer the following questions:

1. Fill in the values in the table when $n = 6$.

2. Fill in the values in the table when $n = 21$.

3. Fill in the values in the table when $n = 60$.

4. Make a new column for 8% interest, both simple and compound. Fill in the values in this new column for $n = 1$ and $n = 2$.

5. How many years does it take for \$1 to double in value at 4% simple interest?

6. How many years does it take for \$1 to double in value at 4% compound interest?

Magic Squares

If we could go back to ancient China or India, we would probably notice some people wearing stone or metal ornaments engraved with an array of numbers similar to those pictured below. Such ornaments were thought to have mystical powers.

8	1	6
3	5	7
4	9	2

16	2	3	13
5	11	10	8
9	7	6	12
4	14	15	1

1	12	7	14
8	13	2	11
10	3	16	5
15	6	9	4

Figure 1

We can at least agree that the type of numerical design is unusual, for if we add the numbers written in any row, column, or diagonal of one of the number arrangements, we will always obtain the same sum. Any such arrangement of numbers in the form of a square so that every column, every row, and each of the two diagonals add up to the same sum is called a *magic square*.

Since their discovery, magic squares have been the source of many mathematical amusements and games. Also, a number of interesting and useful mathematical concepts have been discovered as the result of research on the theory of magic square patterns. Although we no longer attribute special powers to magic squares, we cannot deny their value.

The *order* of a magic square is determined by the number of rows or columns in the square. Thus, a square containing three rows and three columns is said to be a *third order* magic square. The common sum obtained by adding the elements of a row, column, or diagonal is called the *constant* of the square.

It is usually required that magic squares be formed from the consecutive numbers, 1 to n^2, n representing the order of the square. We will confine our discussion to squares of this nature.

Let's try to discover some magic square patterns by studying third order magic squares. Such a square would be made up of the integers 1 to 9, inclusive. We can represent the elements (numbers) of a third order square as shown below.

a	d	g
b	e	h
c	f	i

Figure 2

We can show that there is only one possible value for the constant of such a magic square. The sum of all the elements of the square will be the sum of all the integers from 1 to 9.

$$a + b + c + d + e + f + g + h + i = 1 + 2 + 3 + \cdots + 9$$

From our work with continuous sums, we know that we can find the sum of a series like $1 + 2 + 3 + \cdots + 9$ by averaging the first and last terms of the sum and multiplying that result by the number of terms.

$$1 + 2 + 3 + \cdots + 9 = 9\left(\frac{1 + 9}{2}\right) = 45$$

Since
$$a + b + c = d + e + f = g + h + i,$$
then
$$3(a + b + c) = 45,$$
and
$$a + b + c = 15.$$

Hence the constant for a third order magic square is 15. The same type of reasoning can be used to determine the constant for a magic square of any order.

We can now show that the center number for our third order square must be 5, for

$$(g + e + c) + (d + e + f) + (a + e + i) + (b + e + h) = 60,$$

and by rearrangement we get

$$3e + (a + b + c) + (d + e + f) + (g + h + i) = 60$$
$$3e + 45 = 60$$
$$e = 5$$

We also know that $a + i$, $b + h$, $c + g$, and $d + f$ each equal 10, the possible combinations being 1 and 9, 2 and 8, 3 and 7, and 4 and 6. The 1 and 9 combination cannot occupy a corner position, for there are not three number combinations involving 1 that add up to 15. Also, the integers 8 and 6 must be in the same row or column with 1, for this is the only combination involving 1 (besides 5 and 9) that gives a sum of 15. With this information, it is possible to form all the third order magic squares that are made up of the digits 1 to 9.

EXERCISE SET 11
Working with Magic Squares

1. One third order magic square from the integers 1 to 9 is given on page 34. Construct seven other third order squares from these integers.

2. Show that the constant for a fourth order magic square constructed from the integers 1 to 16 must be 34.

Patterns Involving the Number π

One of the best known mathematical constants is π (the Greek letter, pi). This constant is helpful in expressing many well-known mathematical relationships, such as

$$C = 2\pi r \qquad (C, \text{ the circumference of a circle of radius } r)$$

$$A = \pi r^2 \qquad (A, \text{ the area of a circle of radius } r)$$

$$V = \frac{4}{3}\pi r^3 \qquad (V, \text{ the volume of a sphere of radius } r)$$

The basic definition of π as the ratio of the circumference of a circle to its diameter is very simple, but this constant is more complex than its definition indicates. It is difficult to evaluate π because it is impossible to express *both* the diameter and circumference of a circle as either whole numbers or fractions. Therefore, π cannot be expressed as a whole number or fraction. A number of this nature is said to be *irrational*.

The earliest written records, made on papyrus in Egypt in 1700 B.C., gave directions to find the area of a circle, as follows:

$$A = \left(d - \frac{1}{9}d \right)^2,$$

where A is the area and d is the diameter of the circle. Substituting $2r$ (twice the radius) for d, we get the formula

$$A = \frac{256}{81}r^2.$$

We can thus see that the early Egyptians suggested that π was the ratio $\frac{256}{81}$, which, in decimal form, is $3.16050\ldots$. The difference between this result and the value we use today ($3.14159\ldots$) amounts to about a 2% error.

The great mathematician and scientist of ancient Syracuse, Archimedes (287-212 B.C.), used geometric figures to estimate that π is larger than $3\frac{10}{71}$ but less than $3\frac{1}{7}$, or, expressed in decimals, he estimated π to be between $3.140845\ldots$ and $3.142857\ldots$.

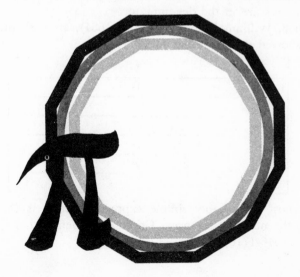

In ancient China, π was expressed by Ch'ang Hong (125 A.D.) as

$$\pi = \sqrt{10} = 3.162 \ldots ..$$

Later, Wang Fan (265 A.D.) estimated π to be

$$\pi = \frac{142}{45} = 3.1555 \ldots,$$

and Ch'ung-chik (470 A.D.) gave a different fraction,

$$\pi = \frac{355}{113} = 3.1415929,$$

which is correct to six decimal places.

Such attempts to find a value of π show that a complete understanding of π was really not known.

The first time π was developed in a regular mathematical pattern was in 1592, when the French mathematician François Vieta showed that π was equal to the following:

$$\pi = 2\left(\frac{1}{\sqrt{\frac{1}{2}} \cdot \sqrt{\frac{1}{2} + \frac{1}{2}\sqrt{\frac{1}{2}}} \cdot \sqrt{\frac{1}{2} + \frac{1}{2}\sqrt{\frac{1}{2} + \frac{1}{2}\sqrt{\frac{1}{2}}} + \cdots}}\right).$$

In 1655, John Wallis, an English mathematician, gave a simpler form:

$$\pi = 4\left(\frac{2 \cdot 4 \cdot 4 \cdot 6 \cdot 6 \cdot 8 \cdot 8 \cdot 10 \cdot 10 \cdot 12 \cdot 12 \cdots}{3 \cdot 3 \cdot 5 \cdot 5 \cdot 7 \cdot 7 \cdot 9 \cdot 9 \cdot 11 \cdot 11 \cdot 13 \cdots}\right).$$

Note that here the even numbers are in the numerator and the odd numbers are in the denominator.

In 1658, William Brouncker (1620?-1684), an Irish mathematician, showed this to be the value of π:

$$\pi = 4 \times \cfrac{1}{1 + \cfrac{1^2}{2 + \cfrac{3^2}{2 + \cfrac{5^2}{2 + \cfrac{7^2}{2 + \cfrac{9^2}{2 + \cdots}}}}}}$$

The most remarkably simple expression for π is this series, bearing the name of the great German mathematician Gottfried Leibniz (1646-1716):

$$\pi = 4\left(1 - \frac{1}{3} + \frac{1}{5} - \frac{1}{7} + \frac{1}{9} - \frac{1}{11} + \frac{1}{13} - \frac{1}{15} + \cdots\right).$$

None of the expressions above are very useful for computing the value of π, because they are either too clumsy or too slow in converging to an accurate result. A better series for computing is

$$\pi = 2\left(1 + \frac{1}{2}\cdot\frac{1}{3} + \frac{1\cdot3}{2\cdot4}\cdot\frac{1}{5} + \frac{1\cdot3\cdot5}{2\cdot4\cdot6}\cdot\frac{1}{7} + \frac{1\cdot3\cdot5\cdot7}{2\cdot4\cdot6\cdot8}\cdot\frac{1}{9} + \cdots\right).$$

In 1717, Abraham Sharp (1651-1742), an English mathematician, used the following series to compute π to 72 decimal places:

$$\pi = 6\frac{1}{\sqrt{3}}\left(1 - \frac{1}{3\cdot3} + \frac{1}{3^2\cdot5} - \frac{1}{3^3\cdot7} + \frac{1}{3^4\cdot9} - \frac{1}{3^5\cdot11} + \cdots\right).$$

In 1873, William Shanks, an English mathematician, published a value for π to 707 decimal places. Then in 1948, John W. French, Jr., of Washington, D. C., and D. F. Ferguson, of Manchester, England, jointly published a corrected and checked value of π to 808 decimal places. They found several errors in Shank's results, beginning with the 528th place.

In the January, 1950, issue of *Mathematical Tables and Other Aids to Computation* is found the value of π calculated to more than 2000 places. This was done, as you might suspect, by our modern-day, high-speed electronic computers. Here is the value of π carried out to 22 decimal places:

$$3.1415926535897932384626\ldots$$

EXERCISE SET 12
Problems with π

1. You might like to compute π to several decimal places yourself by using one of the formulas given in the previous section. You will recognize very soon which formulas are better than others for computing.

2. In tossing coins, you know that a head is just as likely to show as a tail, but in 100 tosses, you wouldn't necessarily come up with exactly 50 heads and 50 tails. However, you could expect a fairly equal distribution of heads and tails as you counted more and more tosses.

The following experiment is quite similar to coin tossing, but instead of tossing coins, you toss sticks. The object of this experiment is to see how close you can come to the value of π.

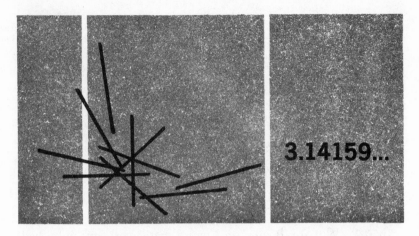

3.14159...

Mark off on a large sheet of paper a series of parallel lines two inches apart. Cut ten sticks, each one inch long; toothpicks will do. Now hold the ten sticks about a foot above the ruled paper and drop them. Count the number of sticks that either touch or cross a line. If you repeat this 100 times, you will have dropped 1000 sticks altogether. Total the number of sticks that have crossed or touched a line and divide this total into 1000, the total number of sticks thrown. This division, theoretically, according to the laws of probability, should give the value of π. Just as in coin tossing you can't expect an exact distribution of heads and tails, so you can't expect to get an exact value of π, but it is fun to see how close you come.

You might increase the number of tosses by having your friends perform the experiment, too, and then combine your results. The general formula to use for any number of tosses is:

$$\frac{\text{Total sticks tossed}}{\text{Total sticks crossing or touching a line}} \quad \text{approximates } \pi.$$

A Backward Glance and a Look to the Future

On the opening pages of this section, you experimented with some interesting number patterns. From this beginning, you found that examination of number patterns could lead to discovery of important number relationships. The next step was to prove the relationships by showing them to be true for *any* number or set of numbers. You were finally led to the topic, infinite series, which represents a small sample of the many unusual number relations that exist. Perhaps the strangest, and most remarkable of all, are those that involve the value of π. It seems unbelievable that a number, starting out to be just a ratio between the circumference and diameter of a circle, should be associated with so many expressions involving number patterns.

The development presented here is often characteristic of all mathematics and science. Through the ages, man has noted fascinating and useful facts, has then used this information to develop concepts and relationships, and has not been satisfied until he has conclusively proven the general nature of the ideas developed. Such efforts have often led to fascinating and remarkable principles.

This is not an outdated art, for the mathematicians and scientists of today and tomorrow will continue to make observations, generalize from these observations, prove their new concepts, and then seek amazing new applications for their ideas.

For Further Reading

If you would like to continue your investigation of the fascinating field of number patterns, here are some books that will give you further information:

BAKST, AARON, *Mathematics, Its Magic and Mastery.* D. Van Nostrand Co., 1952

BALL, W. W. ROUSE, and COXETER, H. S. M., *Mathematical Recreations and Essays.* Macmillan Co., 1947

FRIEND, J. NEWTON, *Numbers: Fun and Facts.* Charles Scribner's Sons, 1954

GARDNER, MARTIN, *Mathematics, Magic and Mystery.* Dover Publications, 1956

JONES, S. I., *Mathematical Clubs* and *Recreations.* S. I. Jones Co., 1940

KRAITCHIK, MAURICE, *Mathematical Recreations.* Dover Publications, 1942

MERRILL, HELEN ABBOT, *Mathematical Excursions.* Dover Publications, 1957

MEYER, JEROME, *Fun with Mathematics.* Dover Publications, 1952

RANSOM, WILLIAM R., *One Hundred Mathematical Curiosities.* J. Weston Walch, 1954

PART III

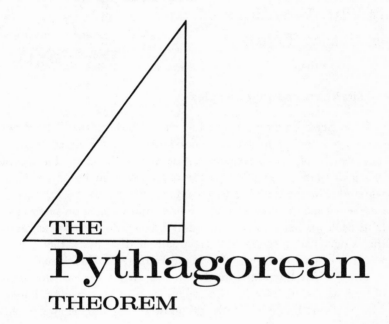

THE
Pythagorean
THEOREM

The Wonders of the Right Triangle

Rope Stretchers and Right Angles

In ancient Egypt every spring the waters of the Nile River overflowed their banks and flooded the land for miles around. The people of Egypt welcomed this annual flooding, for Egypt was a dry land, and this was the only way their crops could be watered. But the flooding was a mixed blessing to the Egyptians, for every time the floods came the lines that marked off one man's land from his neighbor's were washed away. So, every year, after the waters had receded, the Egyptians had to mark their lands again in order to tell just which land belonged to which man.

The ancient Egyptians lived at a time when the measurement of land had to be done in what seems to us to be a crude manner. This was not because the Egyptians were not intelligent people. On the contrary, they were very intelligent — otherwise they could not have made the great pyramids and other wonders of their great civilization. But methods of measurement depend on mathematics, and mathematics had not developed very extensively up to the time of ancient Egypt.

In order to measure their lands, the Egyptians needed to use the right angle, and, to construct a right angle, they used this method: Men known as "rope stretchers" took a rope of a certain length and tied 13 knots at equal intervals on the rope. They then staked the rope on the ground, as shown below, with Stakes 2 and 3 being located at the fourth and eighth knots, and Stake 1 being located where the first and thirteenth knots met. Of course, the rope had to be stretched tight. The angle formed at Stake 2 was then a right angle.

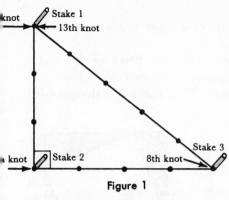

Figure 1

We can see from the triangle formed by the Egyptian rope stretchers that the side of the triangle opposite the right angle measured 5 units, and that the other sides measured 3 and 4 units. The Egyptians were satisfied with their scheme, and it never occurred to them to ask why the relationship of sides in the proportion of 3, 4, 5 resulted in what we now call a right triangle. It was enough for them to know that with this relationship of sides they could get a right angle.

At somewhat the same time, in India, the Hindus also needed to construct right angles, but they went a step farther than the Egyptians. They discovered that in addition to the 3, 4, 5 relationship of sides in a right triangle, there were other relationships that resulted in a right triangle as well. These were as follows:

$$12, 16, 20 \qquad 15, 20, 25 \qquad 5, 12, 13 \qquad 15, 36, 39$$
$$8, 15, 17 \qquad 12, 35, 37$$

But, like the Egyptians, the ancient Hindus never took the time to ask why this relationship held true. If they did ask why, and found the answer to the question, we have never seen a written record of it.

The Right Triangle Question Is Answered

The answer to the question "Why?" was finally found during the sixth century B. C., at about the time of the Golden Age of Greece, presumably by a mathematician and philosopher named Pythagoras (pronounced pĭ-thăg′ō-răs). Whether it was really Pythagoras who found the answer or some other Greek who lived at about the same time is not really known, but at least Pythagoras is given the credit, and his name no doubt will live forever because of the important relationship called the *Pythagorean theorem*.

In every right triangle, the side opposite the right angle is called the *hypotenuse*, and the other two sides are called *legs*. The Pythagorean theorem can then be stated in these words:

"The square of the hypotenuse of a right triangle equals the sum of the squares of the two legs."

It is believed that the earliest geometric proof of this statement was suggested by the tile patterns shown below. Take four right triangles, all of the same size. Then make two designs with these four triangles.

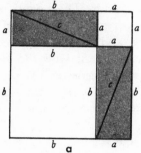
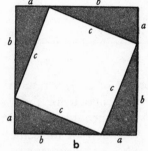

Figure 2

In Figure 2a you can extend the sides to form a large square whose sides are $(a+b)$, the same as the dimensions of the large square in Figure 2b. After removing the four triangles from each of the large squares (see Figure 3), the areas that are left in each figure must be equal.

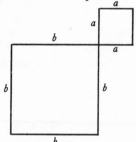
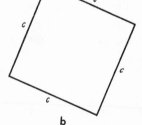

Figure 3

This means that the area of Figure 3a must equal the area of Figure 3b, or,

$$a^2 + b^2 = c^2,$$

which is the statement of the Pythagorean theorem in algebraic form.

The study of this section will give you many interesting examples of how this relationship can help you solve a variety of problems. In effect the Pythagorean theorem says, "If you know any two sides of a right triangle, you can always solve for the third side." You can see how practical this can be by just imagining the many ways that right-angled figures arise in measurement.

The Famous Right Triangle Test

It is easy to test whether or not a triangle forms a right triangle by substituting the values of the lengths of the sides in the formula
$$a^2+b^2=c^2,$$
where c is the largest of the three sides.

Substitution in the formula shows that it is satisfied by the famous 3, 4, 5 triangle.
$$3^2+\ 4^2=5^2$$
$$9\ +16\ =25$$
$$25\ =25$$

Therefore the 3,4,5 triangle is a right triangle with the right angle opposite 5. A scale drawing made with compasses will help to confirm this.

Do the same thing with the three sides 7, 8, and 11.

$$7^2+8^2=$$
$$49+64=113$$

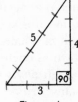

Figure 4

If the Pythagorean theorem is to be satisfied, then the value of 7^2+8^2 should equal 11^2, or 121. But it doesn't. Therefore the triangle is *not* a right triangle. A scale drawing helps to confirm this, although it does not differ greatly from a right triangle. A protractor reading shows the "right angle" to be approximately 96°.

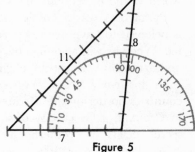

Figure 5

EXERCISE SET 1
Checking for Right Triangles

Which of the following sets of numbers satisfy the Pythagorean theorem? Check your answer with a scale drawing of the triangle.

1. 1, 2, 3	**5.** 6, 8, 10	**8.** 8, 15, 17
2. 2, 3, 4	**6.** 5, 12, 13	**9.** 9, 12, 15
3. 3, 5, 8	**7.** 7, 9, 12	**10.** 15, 20, 25
4. 4, 5, 6		

An Exploration of
Numbers and Geometry

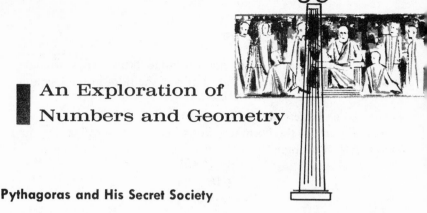

Pythagoras and His Secret Society

We have already considered some aspects of the Pythagorean theorem, but have said little about Pythagoras. Who was this man whose name is associated with such an important theorem?

Not too much is known about Pythagoras. It is believed that he was a pupil of Thales (640 B.C.-550 B.C.), a famous Greek philosopher known as one of the "seven wise men." Neither Thales nor Pythagoras left any writings behind. This doesn't mean that they didn't do any writing, but, if they did, their works were lost. One thing that history is sure of is that without Thales there would not have been a Pythagoras, and without Pythagoras there would not have been a Plato (427 B.C.-347 B.C.). And without Plato the world would be without many wonderful ideas.

Pythagoras was a Greek mathematician and philosopher who is believed to have lived between approximately 569 B.C. and 500 B.C. When he was a young boy, he lived in an atmosphere of culture and art, for it was just the beginning of the Golden Age of Greece.

In his early years, Pythagoras traveled extensively in the countries bordering the Mediterranean Sea. In 509 B.C., he settled in Crotona, a Greek city of southern Italy. Here he organized a secret society or brotherhood. The members of this society were called Pythagoreans. They were most interested in moral reform, but they also had scientific discussions. These led to the development of many important ideas in mathematics and astronomy.

The science of music was another interest of the Pythagoreans. They proved, for example, that if two strings, under the same tension, have lengths in the ratio of 2 to 1, then, when the strings are plucked, the resulting musical notes are an octave apart.

They also were the first to teach that the earth is a globe revolving with other planets around a central sun.

The Pythagoreans seemed to feel that there was something mysterious about numbers — that numbers were the foundation of life. They even attached significance to specific numbers. The following illustrate a few cases:

    1 stood for reason because it was unchangeable.

    2 stood for masculine.

    3 stood for feminine.

    4 stood for justice because it was the first product of equals $(2 \times 2 = 4)$.

    5 stood for marriage — the union of the first masculine number and the first feminine number $(2 + 3 = 5)$.

The Pythagoreans also associated numbers with geometric patterns. The sum of consecutive odd numbers provides one example of this association. The geometric pattern is shown here in the form of dots.

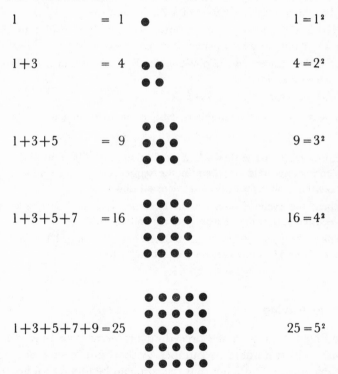

$$1 = 1 \qquad 1 = 1^2$$

$$1 + 3 = 4 \qquad 4 = 2^2$$

$$1 + 3 + 5 = 9 \qquad 9 = 3^2$$

$$1 + 3 + 5 + 7 = 16 \qquad 16 = 4^2$$

$$1 + 3 + 5 + 7 + 9 = 25 \qquad 25 = 5^2$$

The addition of each consecutive odd number produces a new sum which can be represented by squared patterns of dots.

This led the Pythagoreans to seek other relationships. Could they, for example, find a square which was the sum of two other squares? One solution can be seen from the above array, namely,

$$3^2 + 4^2 = 5^2.$$

They discovered other similar relationships by adding consecutive odd numbers until they reached an odd number that was a square of a whole number. A number that is the square of a whole number is called a *perfect square*.

The following exercises will enable you to discover some of the same relationships that fascinated the Pythagoreans.

EXERCISE SET 2
Number Patterns and the Pythagorean Theorem

1. What is the sum of the consecutive odd numbers from 1 to 23? Notice that the answer is a square of a whole number.

2. What is the sum of the consecutive odd numbers from 1 to 25? This answer should also be a perfect square.

3. See if the answers to the questions in Exercises 1 and 2 lead you to the Pythagorean relation,

$$a^2 + b^2 = c^2,$$

where a, b, and c are whole numbers. Notice the arrangement:

$$(1 + 3 + 5 + \ldots + 23) + (25) = (1 + 3 + 5 + \ldots + 25).$$

Each expression in parentheses is a perfect square. The dots indicate that all consecutive odd numbers in the sequence are to be added. It would take too much space to write them all down.

4. Repeat the cycle of questions in Exercises 1, 2, and 3 by adding the consecutive odd numbers from 1 to 47 and 1 to 49.

5. Repeat the cycle of questions in Exercises 1, 2, and 3 by adding the consecutive odd numbers from 1 to 79 and 1 to 81.

Short Cuts to Adding

In finding the sums in the preceding problems, you may have laboriously added number by number. A short cut to such adding can be developed by noticing that the sum can be found as follows:

Consider $1 + 3 + 5 + 7 + 9$. Because these 5 numbers are equally spaced, it perhaps sounds reasonable to say that their sum can be obtained by finding the average of the first and last numbers and then multiplying this average by the number of terms being added. The proof of this idea is often a part of the high school advanced algebra course.

In the above example the average of the first and the last terms is $\dfrac{1+9}{2}$, and this expression, multiplied by 5 (the number of terms being added), gives the desired sum. Thus,

$$1+3+5+7+9 = \frac{1+9}{2} \times 5 = \left(\frac{10}{2}\right)(5) = (5)\ (5) = 25.$$

Notice that this sum is a perfect square.

Another way to add the series is to reverse the terms in the series and add in the following fashion:

$$
\begin{array}{r}
1+\ 3+\ 5+\ 7+\ 9 \\
9+\ 7+\ 5+\ 3+\ 1 \\
\hline
10+10+10+10+10.
\end{array}
$$

The sum of these tens is 5(10), but this is twice the answer desired because the series was added to itself. Therefore,

$$1+3+5+7+9 = \frac{5(10)}{2} = (5)\ (5) = 25.$$

Notice that this sum is a perfect square and agrees with the answer obtained above.

EXERCISE SET 3
Short-Cut Practice

By means of either of the methods shown above, find the sum of each of the following series:

1. $1+3+5+\ldots+23=$
2. $1+3+5+\ldots+23+25=$
3. $1+3+5+\ldots+47=$
4. $1+3+5+\ldots+47+49=$
5. $1+3+5+\ldots+79=$
6. $1+3+5+\ldots+79+81=$
7. $1+3+5+\ldots+119=$
8. $1+3+5+\ldots+119+121=$
9. $1+3+5+\ldots+167=$
10. $1+3+5+\ldots+167+169=$
11. $1+3+5+\ldots+223=$
12. $1+3+5+\ldots+223+225=$

Exploring an Interesting Set of Numbers

When three whole numbers, a, b, and c, satisfy the relation $a^2+b^2=c^2$, they are called *Pythagorean numbers*. A big question now arises. Is the technique used in Exercise Set 2 going to give all possible combinations of whole numbers that satisfy the Pythagorean formula? The answer is "No." The numbers 8, 15, and 17 would not have been found by the above method, yet they satisfy the formula.

$$8^2 + 15^2 = 17^2$$
$$64 + 225 = 289$$
$$289 = 289$$

Pythagoras was able to develop the following rule for finding three numbers that satisfy the relation $a^2 + b^2 = c^2$. These three numbers can be represented by

$$m, \quad \frac{1}{2}(m^2 - 1), \quad \text{and } \frac{1}{2}(m^2 + 1),$$

where m can be assigned the value of any odd whole number. These three are Pythagorean numbers because they satisfy the formula

$$a^2 + b^2 = c^2,$$

which means that the following expression is true for all odd whole number values of m:

$$m^2 + \left\{ \frac{1}{2}(m^2 - 1) \right\}^2 = \left\{ \frac{1}{2}(m^2 + 1) \right\}^2.$$

For example, let $m = 3$. Then the Pythagorean numbers are:

$$3, \frac{1}{2}(3^2 - 1), \text{ and } \frac{1}{2}(3^2 + 1),$$

and these simplify to

$$3, \frac{1}{2}(9 - 1), \text{ and } \frac{1}{2}(9 + 1),$$

or

$$3, \frac{1}{2}(8), \quad \text{and } \frac{1}{2}(10),$$

and finally,

$$3, \quad 4, \quad \text{and } \quad 5.$$

EXERCISE SET 4
Pythagorean Numbers

1. Prove that this formula is true for all values of m:

$$m^2 + \left\{ \frac{1}{2}(m^2 - 1) \right\}^2 = \left\{ \frac{1}{2}(m^2 + 1) \right\}^2$$

2. Find the Pythagorean numbers for $m=5$; for $m=7$; for $m=9$; for $m=11$.

3. What happens in the formula when m is an even number? By trying an even number, you should find that you no longer have whole numbers, but the fractional values still satisfy the Pythagorean theorem.

4. Prove that the following expression is true for all values of a, b, c, and d:
$$(ac-bd)^2+(ad+bc)^2=(a^2+b^2)(c^2+d^2).$$

5. In Exercise 4, substitute m for both a and c, n for both b and d. Show that the expression becomes
$$(m^2-n^2)^2+(2mn)^2=(m^2+n^2)^2.$$
This is still a more general method of obtaining Pythagorean numbers, if m and n are assigned arbitrary values of positive whole numbers.

6. Using the expression in Exercise 5, find Pythagorean numbers by letting

(a) $m=2$, $n=1$ (d) $m=4$, $n=1$
(b) $m=3$, $n=1$ (e) $m=4$, $n=2$
(c) $m=3$, $n=2$ (f) $m=4$, $n=3$.

7. Show that when $n=1$, the expression in Exercise 5 can be changed to the one credited to Pythagoras:
$$m^2+\left\{\frac{1}{2}(m^2-1)\right\}^2=\left\{\frac{1}{2}(m^2+1)\right\}^2.$$

A Broader Horizon for the Pythagorean Numbers

One of the common characteristics of mathematicians is that they are always seeking ways to extend an idea. The Pythagoreans looked for whole number solutions of the expression
$$a^2+b^2=c^2.$$
It is quite natural to go a step farther and seek whole number solutions of the sum of three squares equaling a square,
$$a^2+b^2+c^2=r^2.$$
Books on the theory of numbers treat such problems in great detail. It may interest you at this point to see the list of all possible solutions for certain values of r. You may wish to verify that these are correct solutions:
$$r=3$$
$$a=1, \quad b=2, \quad c=2$$
$$r=5$$
no values for a, b, c.

$$r = 7$$
$$a = 2, \quad b = 3, \quad c = 6$$

$$r = 9$$
$$a = 1, \quad b = 4, \quad c = 8$$
$$a = 3, \quad b = 6, \quad c = 6$$
$$a = 4, \quad b = 4, \quad c = 7$$

$$r = 11$$
$$a = 2, \quad b = 6, \quad c = 9$$
$$a = 6, \quad b = 6, \quad c = 7$$

A Strange New Number Is Discovered

The discussion of Pythagorean numbers may lead you to believe that whole number relationships were the chief concern of the Greeks, but this was not entirely so. It is true that Pythagoras and his followers were fascinated by the relationships of whole numbers, or integers, as they are also called. The Pythagoreans thought that everything in the universe could be explained by the integers. They acknowledged defeat, however, when they found that it was impossible to find two whole numbers such that the square of one of them is equal to twice the square of the other one. The problem they tried to solve with whole numbers was expressed by the formula:

$$m^2 = n^2 + n^2.$$

This formula can be changed to any one of the following forms:

$$m^2 = 2n^2, \text{ or } \frac{m^2}{n^2} = 2, \text{ or } \left(\frac{m}{n}\right)\left(\frac{m}{n}\right) = 2.$$

A number that can be multiplied by itself to give 2 is called the "square root of two" (represented by the symbol $\sqrt{2}$).

The Pythagoreans discovered that no whole number values of m and n would make the above formulas true. Another way to interpret their conclusion is that the "square root of two" cannot be expressed as a ratio of two whole numbers. Any number that can be expressed as the ratio of two whole numbers is called a rational number. Thus, $\sqrt{2}$ is not a rational number, and so a new name was invented for it — "irrational," that is, not a ratio of two whole numbers. A few examples of rational numbers are

$\dfrac{1}{2}$, $\dfrac{2}{3}$, $\dfrac{5}{4}$, and so on. A few examples of irrational numbers are $\sqrt{2}$, $\sqrt{3}$, $\sqrt{20}$, and so on.

As a result of the strange behavior of the irrational numbers, the Pythagoreans tried to limit their discussions of mathematics to rational numbers. The story is told that the Pythagoreans were so upset about numbers which could not be expressed as a whole number divided by another whole number that they tried to keep it a secret. Legend has it that one of the Pythagoreans, Hippasus, was drowned by his fellow brothers for telling the secret to outsiders.

Geometry and the Pythagorean Theorem

Although the Pythagoreans were disturbed by irrational numbers, they did make use of them in geometric situations. In fact, by considering a, b, and c as the three sides of a right triangle, they proved that the formula

$$a^2 + b^2 = c^2$$

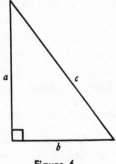

Figure 6

was true no matter what the lengths of the sides might be, thus firmly establishing the famous theorem, "The square on the hypotenuse of a right triangle equals the sum of the squares of the legs."

If any two numbers, rational or irrational, are given as the lengths of two sides of a right triangle, the third side is determined by the above formula.

Hundreds of proofs of the Pythagorean theorem have been developed through the centuries. *The Pythagorean Theorem*, a book by E. S. Loomis, contains a classified collection of over 370 different proofs of the theorem.

In about 300 B.C. Euclid, a famous Greek mathematician, recorded a proof of the theorem. Euclid is remembered for his compilation of the elementary parts of geometry in thirteen booklets, known now as Euclid's *Elements*. These *Elements* have been the basis for all subsequent books on elementary geometry.

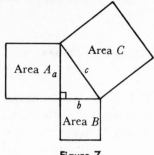

Figure 7

Euclid's proof of the theorem uses the diagram in which squares are constructed on each side of the right triangle.

He then proceeds to prove that Area A+Area B=Area C. Since a^2 = Area A, b^2 = Area B, and c^2 = Area C, it is true that $a^2+b^2=c^2$.

Three simple proofs that combine the tools of algebra and geometry follow:

Method I. Cut out four identical right triangles:

  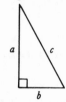 

Figure 8

Place these triangles as in Figure 9.

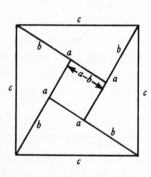

Figure 9

Notice that a large square is formed of side c, and its area is therefore c^2. But this area is made up of the center square plus the four triangles. The center square has an area of $(a-b)^2$. The four triangles have a total area of $4\left(\frac{1}{2}ab\right)$. Therefore, the area of the large square, c^2, equals the sum of the areas of the center square, $(a-b)^2$, and the four triangles, $4\left(\frac{1}{2}ab\right)$. Or,

$$c^2 = (a-b)^2 + 4\left(\frac{1}{2}ab\right),$$

which simplifies to

$$c^2 = (a^2 - 2ab + b^2) + 2ab,$$

or finally to

$$c^2 = a^2 + b^2.$$

Method II. Take the same four triangles shown in Method I and arrange them as in Figure 10.

Then, on the basis of areas, you have the relationship,

$$(a+b)^2 = 4\left(\frac{1}{2}ab\right) + c^2,$$

which simplifies to

$$(a^2 + 2ab + b^2) = 2ab + c^2,$$

or finally to

$$a^2 + b^2 = c^2.$$

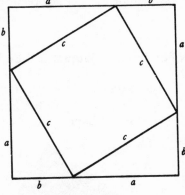

Figure 10

Method III. Start with the right triangle *ABC* and drop perpendicular *CD* from the vertex of the right angle to the hypotenuse.

Two triangles are similar (have the same shape) if two angles of one triangle are equal to two angles of the other. Triangles *CDB* and *ACB* both contain angle *B*, and both contain right angles. Thus they are similar. Triangles *ADC* and *ACB* are also similar, for right angle *ADC* is equal to right angle *ACB*, and angle *A* is contained in each triangle. Since the corresponding sides of similar triangles

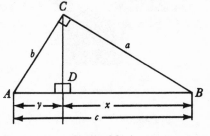

Figure 11

are proportional, the sides of the triangles in the above figure form the following two proportions:

$$(1) \quad \frac{x}{a} = \frac{a}{c} \quad \text{and} \quad (2) \quad \frac{y}{b} = \frac{b}{c}.$$

But it is also true that

$$(3) \quad x + y = c.$$

From equations (1) and (2), you can find that $x = \dfrac{a^2}{c}$ and $y = \dfrac{b^2}{c}$.

Substituting these expressions for *x* and *y* in equation (3), you can obtain

$$\frac{a^2}{c} + \frac{b^2}{c} = c,$$

and clearing fractions by multiplying both sides by c gives the formula for the Pythagorean theorem,

$$a^2+b^2=c^2.$$

A Fascinating Geometric Side Trip

The geometric interpretation of the Pythagorean theorem, "the sum of the squares on the legs of a right triangle is equal to the square on the hypotenuse," actually shows squares constructed on the sides.

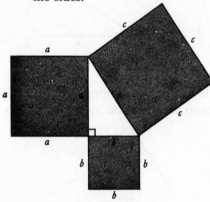

Figure 12

It is interesting to note that other figures placed on these sides also have the same area relationship as long as the three figures are similar to each other. The reason for this is that when any two figures are similar, their areas are proportional to the squares of a corresponding dimension. For example, if two triangles are similar, their areas are proportional to the squares of a pair of corresponding sides.

Example: Construct equilateral triangles (triangles having all sides equal) on the three sides, a, b, and c, of the right triangle.

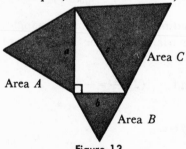

Area A

Area C

Area B

Figure 13

Since all equilateral triangles are similar, the areas of the triangles in the above figure are proportional to the squares of the corresponding sides. In other words,

$$\frac{\text{Area } A}{\text{Area } C}=\frac{a^2}{c^2} \quad \text{and} \quad \frac{\text{Area } B}{\text{Area } C}=\frac{b^2}{c^2}.$$

Remembering that $a^2+b^2=c^2$ is a true relationship because we have a right triangle given, we can divide both sides of this expression by c^2 to obtain

$$\frac{a^2}{c^2}+\frac{b^2}{c^2}=1.$$

But this is the same as

$$\frac{\text{Area } A}{\text{Area } C} + \frac{\text{Area } B}{\text{Area } C} = 1,$$

and clearing fractions results in the statement we wanted to prove:

$$\text{Area } A + \text{Area } B = \text{Area } C.$$

Thus the equilateral triangle on the hypotenuse of a right triangle is equal to the sum of the equilateral triangles on the legs.

EXERCISE SET 5
Some Interesting Area Problems

In Exercises 1-8 prove that Area A + Area B = Area C. You need to use only the fact that in a particular problem the figures constructed on the three sides of the right triangle are similar to each other.

1. Semicircles:

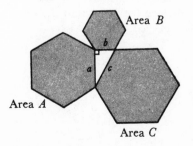

3. Regular octagons (eight-sided figures with all sides equal and all angles equal):

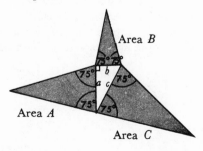

2. Regular hexagons (six-sided figures with all sides equal and all angles equal):

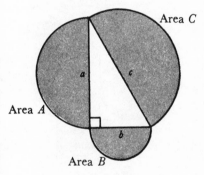

4. Isosceles triangles with base angles of 75° (two sides equal):

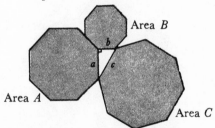

5. Similar triangles:

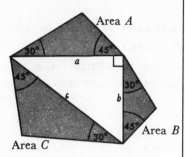

Area A

Area B

Area C

7. Arches:

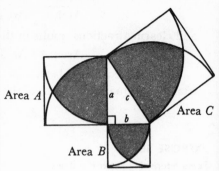

Area A

Area B

Area C

6. Quarter circles:

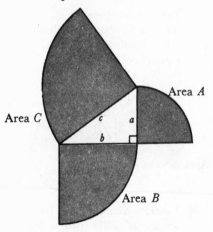

Area A

Area B

Area C

8. Circles:

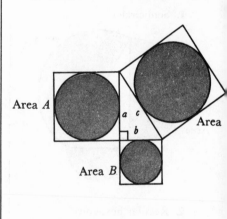

Area A

Area B

Area

9. Prove that the same amount of water will flow through two pipes, one 3 inches in diameter, the other 4 inches in diameter, as will flow through a pipe 5 inches in diameter, provided the water is at the same pressure in all pipes and that friction is neglected.

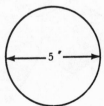

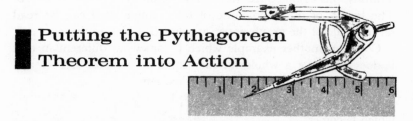

Putting the Pythagorean Theorem into Action

Finding the Hypotenuse: Compasses and Ruler Department

So far you have done quite a bit with the Pythagorean theorem without doing the kind of computing that is usually associated with the formula — namely, finding one side when two other sides are given.

If the two legs a and b are given, the hypotenuse c can be obtained by solving for c in the formula

$$c^2 = a^2 + b^2.$$

All that needs to be done to find c is to take a square root:

$$c = \sqrt{a^2 + b^2}.$$

There are a number of ways to obtain the answer to such a problem. The first way we shall consider is very simple, and requires only graph paper and compasses. The metric scale or graph paper scaled in tenths of an inch is best because answers can be read decimally.

Suppose we start with the problem of finding the hypotenuse when one leg is 3 units and the other leg is 4 units. We know from the discussion of Pythagorean numbers that the hypotenuse should measure 5 units because $5^2 = 3^2 + 4^2$.

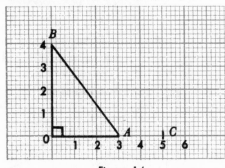

Figure 14

Let's use the graph paper to verify this. First lay off vertical and horizontal lines on the graph paper to establish the right angle. Then lay off $OB = 4$ units on the vertical axis and $OA = 3$ units on the horizontal axis. Open the compasses to measure AB, the hypotenuse. Transfer this distance AB to the horizontal axis so that $OC = AB$. The length of the hypotenuse of the triangle ACB can be read on the scale of the graph paper at point C, which in this case is 5.

Consider another example which is somewhat different in that it does not have a whole number solution. Suppose $OA = 1$ and $OB = 1$. Then, by the Pythagorean theorem,

$$\overline{AB}^2 = \overline{OA}^2 + \overline{OB}^2,$$

which becomes,

$$\overline{AB}^2 = 1^2 + 1^2,$$
$$= 1 + 1,$$
$$= 2.$$

Therefore, $AB = \sqrt{2}$.

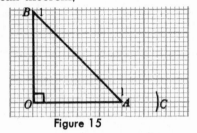

Figure 15

By setting $AB = OC$ on the axis, you can read the approximate length of OC as 1.41 (an estimate to the nearest hundredth of a unit).

EXERCISE SET 6
Finding the Hypotenuse

For Exercises 1-4, use graph paper to find the length of the hypotenuse of a right triangle having the legs given below. In each case, give the number whose square root you have found.

1.   Leg = 2 units

Leg = 1 unit

3.   Leg = 3 units

Leg = 1 unit

2.   Leg = 2 units

Leg = 2 units

4.   Leg = 3 units

Leg = 2 units

5. Use the Pythagorean theorem to prove that if the legs of a right triangle are 1 and $\sqrt{N}$, then the hypotenuse is $\sqrt{N+1}$.

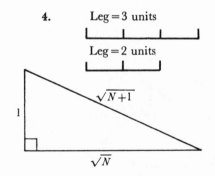

114

If N is a whole number, this says that you can construct $\sqrt{N+1}$ if you know $\sqrt{N}$ and 1. Using this principle, verify the correctness of the figure at the right.

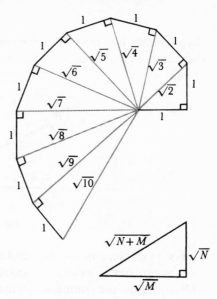

6. A more general relation can be obtained by having a right triangle whose legs are $\sqrt{N}$ and $\sqrt{M}$.

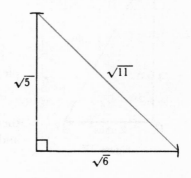

Prove that the hypotenuse has the value $\sqrt{N+M}$.

Apply this principle to find $\sqrt{5}$ by using lengths of $\sqrt{2}$ and $\sqrt{3}$ units.

7. You can use the various lengths for the square roots in the spiral figure of Exercise 5 to construct new square roots not shown in this figure. For example, the square root of 11 units can be constructed from the square root of 5 and the square root of 6 by using the figure below.

By this method, construct:

    (a) $\sqrt{13}$ or $\sqrt{9+4}$

    (b) $\sqrt{15}$ or $\sqrt{8+7}$

    (c) $\sqrt{20}$ or $\sqrt{10+10}$

Check your accuracy in part (c) and notice if two lengths of the square root of 5 are the same as the square root of 20; that is, is this true: $\sqrt{20} = 2\sqrt{5}$?

Finding a Leg: Compasses and Ruler Department

The previous problems may have left you with the impression that the only problems you can solve are those in which the two legs of the right triangles are known and you are to find the hypotenuse. But this is not the case. A leg of a right triangle can be found by construction when the hypotenuse and other leg are given.

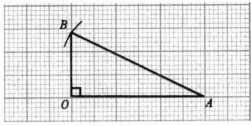

Figure 16

For example, suppose OA and AB are given. Then lay off OA on the horizontal axis of the graph paper. With A as center, and AB as radius, use compasses to mark point B on the vertical axis. OB is the desired length.

The Pythagorean theorem will apply to this situation because triangle OAB is a right triangle. Therefore,

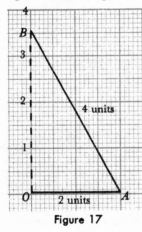

Figure 17

$$\overline{AB}^2 = \overline{OB}^2 + \overline{OA}^2,$$

or

$$\overline{OB}^2 = \overline{AB}^2 - \overline{OA}^2,$$

or

$$OB = \sqrt{\overline{AB}^2 - \overline{OA}^2}.$$

If you do this for a specific example, such as,

the leg $OA = 2$ units,

the hypotenuse $AB = 4$ units,

then OB will measure between 3.4 and 3.5, estimated as 3.46.

EXERCISE SET 7
Finding a Leg

Use graph paper to find the length of a leg of a right triangle having the hypotenuse and the other leg given. In each case, list the number whose square root you have found.

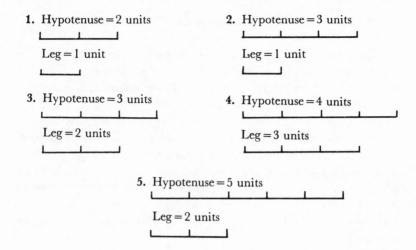

1. Hypotenuse = 2 units

Leg = 1 unit

2. Hypotenuse = 3 units

Leg = 1 unit

3. Hypotenuse = 3 units

Leg = 2 units

4. Hypotenuse = 4 units

Leg = 3 units

5. Hypotenuse = 5 units

Leg = 2 units

Square Roots from Educated Guesses

If we wish to find the value of the square root of any number, we are trying to find a number such that, when it is multiplied by itself, it gives the original number. If we wanted to find the square root of 2, we could use the method of Exercise Set 6 and construct a right triangle with each leg 1 unit in length. The length of the hypotenuse would then be the square root of two units. Instead of using a geometric construction, we could obtain an answer by a trial-and-error method.

First try 1 for the answer. This is too small because the product of 1 times 1 is 1, which is less than the desired product, 2.

Now try 2 for the answer. This is too large because 2 times 2 equals 4, and we are trying to get a product which equals 2.

We now know that our answer must lie between 1 and 2.

Next try 1.5 for the answer. But $1.5 \times 1.5 = 2.25$. This is too large, but the product is fairly close to 2.

So now try 1.4 for the answer. This is getting closer because $1.4 \times 1.4 = 1.96$, and this is only .04 less than the correct product, 2. We now know that the answer lies somewhere between 1.4 and 1.5.

Try 1.41 next. Then $1.41 \times 1.41 = 1.9881$. The square of 1.41 is less than 2.

Try 1.42 next. Then $1.42 \times 1.42 = 2.0164$, which shows that the square of 1.42 is greater than 2.

We now know that the square root of 2 lies between 1.41 and 1.42.

This process may be continued to obtain the accuracy to as many decimal places as desired. In fact, the value can be located between 1.414 and 1.415. As a check, we might square both of these to see if 2 lies between these squares.

This is not a very impressive method, for it is quite tedious. There is too much trial and error and slow computation, even though the idea behind it is quite important. Perhaps you have wondered if there isn't a quicker and surer way. In fact, there are several other ways to find square roots. The next two sections give an explanation of two more methods and the section beyond that shows how you can use tables to find approximate square root values.

An Interesting Square Root Formula

The method of estimating the square root of a number, then squaring it to see whether it is too large or too small, can be modified to eliminate the guesswork. Suppose you wanted to obtain an approximation to $\sqrt{2}$. If the first estimate were 1, then the second estimate would be 2 divided by 1, which is 2. We now suggest that the next fair estimate to try would be the average of 1 and 2. This average is $\frac{1+2}{2}$, which equals $\frac{3}{2}$. Using $\frac{3}{2}$ (which is on the high side) as the new estimate, you can divide 2 by $\frac{3}{2}$ to get $2 \div \frac{3}{2} = \frac{4}{3}$, a corresponding estimate on the low side. The average of these two will give a still better estimate:

$$\frac{\frac{3}{2}+\frac{4}{3}}{2} = \left(\frac{9}{6}+\frac{8}{6}\right) \div 2 = \frac{9+8}{12} = \frac{17}{12} = 1.416$$

This method can be continued as long as you wish. Each step will give greater accuracy than the step before. This type of solution is particularly useful for modern electronic computing machines because it involves only the very simple steps of feeding instructions to the machine that tell it to perform a sequence of arithmetic operations — first a division, then an addition, and finally another division. The cycle is repeated until the desired accuracy is reached.

The steps shown above can be used to compute the square root of any number N. Suppose you select a first estimate of $\sqrt{N}$. Call this x_1. Then divide N by x_1, obtaining $\frac{N}{x_1}$. Now average x_1 and $\frac{N}{x_1}$ to obtain the improved estimate x_2. The formula can be written as:

$$x_2 = \frac{x_1 + \dfrac{N}{x_1}}{2}$$

This formula seems very reasonable because it essentially says, "I am looking for a number, which, when multiplied by itself, gives N. If x_1 is a good guess, but too large, then $\frac{N}{x_1}$ will be too small, but the average of these ought to give a better estimate than either of them." This is exactly what x_2 is — the average of x_1 and $\frac{N}{x_1}$.

This type of formula is called a *recursion formula* because you can take the improved answer x_2 and substitute it in the formula as x_1 to obtain a still better answer, and this process may *recur* over and over again until the desired accuracy is obtained.

Let's do the problem of finding the square root of 2 once again, using decimal values throughout.

Start with $x_1 = 1$. Then

$$x_2 = \frac{1 + \dfrac{2}{1}}{2} = \frac{1+2}{2} = \frac{3}{2} = 1.5.$$

Now use $x_1 = 1.5$. Then

$$x_2 = \frac{1.5 + \dfrac{2}{1.5}}{2} = \frac{1.5 + 1.33}{2} = \frac{2.83}{2} = 1.415.$$

Now use $x_1 = 1.415$. Then

$$x_2 = \frac{1.415 + \dfrac{2}{1.415}}{2} = \frac{1.415 + 1.41342}{2} = \frac{2.82842}{2} = 1.41421$$

Now use $x_1 = 1.41421$. Then

$$x_2 = \frac{1.41421 + \dfrac{2}{1.41421}}{2} = \frac{1.41421 + 1.4142171}{2} = 1.4142135,$$

and since x_1 and x_2 now agree with each other to the nearest hundred thousandth, you can stop unless you want still greater accuracy.

EXERCISE SET 8
Computing Square Roots

Use the recursion formula to compute the square root of the numbers in Exercises 1-3 to the nearest hundredth.

1. $\sqrt{3}$ **2.** $\sqrt{5}$ **3.** $\sqrt{15}$

4. Multiply the answer obtained in Exercise 1 by the answer obtained in Exercise 2 and see if the result is the same, to the nearest hundredth, as the answer in Exercise 3; that is, does $\sqrt{3}$ times $\sqrt{5}$ equal $\sqrt{15}$?

5. An interesting relationship is that

$$(\sqrt{3} + \sqrt{2})(\sqrt{3} - \sqrt{2}) = 1.$$

This can be proved by using the fact that

$$(a+b)(a-b) = a^2 - b^2,$$

for $(\sqrt{3} + \sqrt{2})(\sqrt{3} - \sqrt{2}) = (\sqrt{3})^2 - (\sqrt{2})^2 = 3 - 2 = 1.$

Compute $\sqrt{3} + \sqrt{2}$ and compute $\sqrt{3} - \sqrt{2}$. Is the product of these two equal to 1?

6. Without computing square roots, but using only the identity equation,

$$(a+b)(a-b) = a^2 - b^2,$$

find the value of $(\sqrt{5} + \sqrt{3})(\sqrt{5} - \sqrt{3})$. In general, can you say $(\sqrt{N} + \sqrt{M})(\sqrt{N} - \sqrt{M}) = N - M$ for all values of N and M?

From Areas to Square Roots in One Easy Lesson

Another way to compute square roots can be illustrated by considering a square whose area is 1,444 square inches. A square root computation would arise from a problem like the following:

"What is the length of the side of a square whose area is 1,444 square inches?"

In Figure 18, *MNOP* represents the square having a 1,444 square inch area. In order to find the length of the side, we divide *MNOP* into four parts. We want one part to be the largest square having dimensions that contain only one digit that is not zero that could be placed inside the original square. A square with 30-inch sides has an area of 900 square inches. A 40-inch

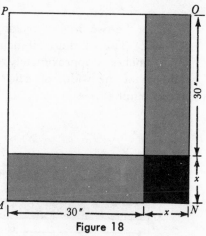

Figure 18

square has an area of 1,600 square inches, which exceeds the original area. We thus choose 30 inches as the dimensions of the white square in Figure 18. We then divide the remaining area into three parts: a black square whose sides are represented by *x*, and two rectangles having the dimensions 30 inches by *x* inches.

Deducting the 900 square inch area of the white square from the original area of 1,444 square inches leaves 544 square inches for the combined area of the two rectangles and the small black square. For a better picture of this area, place the rectangles. end to end and the black square at the right end. The three figures form a new rectangle, *ABCD*, having a total area of 544 square inches.

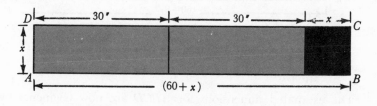

Figure 19

We now wish to estimate the width, *x*, of *ABCD*. Since the black square is small compared to the two original rectangles, the approximate width of *ABCD* is found by dividing the combined length of the two original rectangles into the 544 square inch area of *ABCD*. Thus, 2 times 30 inches, or 60 inches, divided into 544 square inches is approximately 9 inches. If 9 inches is used as an estimate of the width of *ABCD*, a strip 69 inches by 9 inches would result.

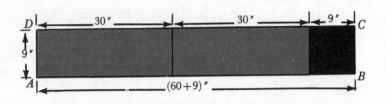

Figure 20

The area of this strip is 69 inches $\times$ 9 inches, or 621 square inches. This is greater than 544 square inches, so the estimate for the width must be reduced from 9 inches to 8 inches.

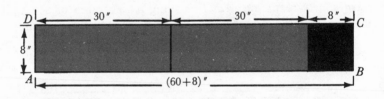

Figure 21

The estimated dimensions for *ABCD* are now 68 inches by 8 inches, giving the desired area of 544 square inches. We thus know that the side of the original square is 30+8 inches, or 38 inches.

The above area method for finding a square root can become quite tedious. However, it is possible to transform the area concepts of the method into simple steps of arithmetic. Compare the following arithmetic steps to the steps of the area method.

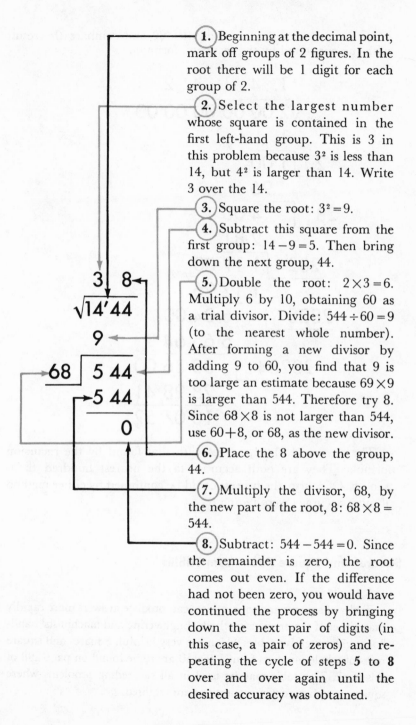

1. Beginning at the decimal point, mark off groups of 2 figures. In the root there will be 1 digit for each group of 2.

2. Select the largest number whose square is contained in the first left-hand group. This is 3 in this problem because 3^2 is less than 14, but 4^2 is larger than 14. Write 3 over the 14.

3. Square the root: $3^2 = 9$.

4. Subtract this square from the first group: $14 - 9 = 5$. Then bring down the next group, 44.

5. Double the root: $2 \times 3 = 6$. Multiply 6 by 10, obtaining 60 as a trial divisor. Divide: $544 \div 60 = 9$ (to the nearest whole number). After forming a new divisor by adding 9 to 60, you find that 9 is too large an estimate because 69×9 is larger than 544. Therefore try 8. Since 68×8 is not larger than 544, use $60 + 8$, or 68, as the new divisor.

6. Place the 8 above the group, 44.

7. Multiply the divisor, 68, by the new part of the root, 8: $68 \times 8 = 544$.

8. Subtract: $544 - 544 = 0$. Since the remainder is zero, the root comes out even. If the difference had not been zero, you would have continued the process by bringing down the next pair of digits (in this case, a pair of zeros) and repeating the cycle of steps **5** to **8** over and over again until the desired accuracy was obtained.

Let's use this method to compute $\sqrt{2}$ and compare the result with the one found by the recursion formula.

$$
\begin{array}{r}
1.\ 4\ \ 1\ \ 4\ \ 2\ \ 1 \\
\hline
\sqrt{2.00'00'00'00'00'}
\end{array}
$$

```
           1. 4  1  4  2  1
        √ 2.00'00'00'00'00'
          1
    24 | 1 00
          96
   281 |    4 00
          2 81
  2824 | 1 19 00
        1 12 96
 28282 |    6 04 00
          5 65 64
282841 |    38 36 00
          28 28 41
          10 07 59
```

The answer, 1.41421, agrees with that found by the recursion formula. They are both accurate to the nearest hundred thousandth. Of course, the process could be continued by either method if more accuracy were required.

Squares and Square Roots from Tables

Tables of squares and square roots can produce answers more rapidly than either of the previous methods. Engineering and machinists' handbooks give extensive tables which are very helpful. Squares and square roots of whole numbers from 1 to 150 are to be found on page 298 of this book. This table may be used in all succeeding problems where square or square root computations are required.

Square Root Applications

In Exercises 1-7, find the value of the unknown side of each right triangle. Use the table of squares and square roots if you wish.

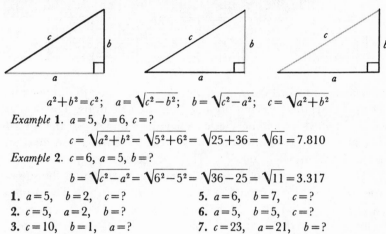

$$a^2+b^2=c^2; \quad a=\sqrt{c^2-b^2}; \quad b=\sqrt{c^2-a^2}; \quad c=\sqrt{a^2+b^2}$$

Example 1. $a=5$, $b=6$, $c=?$
$$c=\sqrt{a^2+b^2}=\sqrt{5^2+6^2}=\sqrt{25+36}=\sqrt{61}=7.810$$

Example 2. $c=6$, $a=5$, $b=?$
$$b=\sqrt{c^2-a^2}=\sqrt{6^2-5^2}=\sqrt{36-25}=\sqrt{11}=3.317$$

1. $a=5$, $b=2$, $c=?$
2. $c=5$, $a=2$, $b=?$
3. $c=10$, $b=1$, $a=?$
4. $c=8$, $a=4$, $b=?$
5. $a=6$, $b=7$, $c=?$
6. $a=5$, $b=5$, $c=?$
7. $c=23$, $a=21$, $b=?$

8. Find the answers to the above problems if each of the sides is doubled; tripled; 10 times as large; 100 times as large.

A Slide Rule for the Pythagorean Theorem

Mathematicians are always searching for ways to shorten computations. You are perhaps familiar with an instrument used for this purpose, the slide rule. You can make a cardboard slide rule to solve the Pythagorean theorem in the following simple way: Take two strips of cardboard, each approximately $2'' \times 18''$, and paste a strip of graph paper on each. Use either the metric scale or tenths to the inch. The metric scale is shown here in Figure 22. Let 1 millimeter be the basic unit on the slide rule. Label the beginning of the scale with a zero. Next mark a 1 one millimeter to the right of the 0-mark to represent 1^2. Then mark a 2 four millimeters to the right of the 0-mark to represent 2^2. The 3 is placed at 9 millimeters from 0; the 4 at 16 millimeters from 0; and so on.

Intermediate marks may be placed by squaring the desired value. For example, 1.5 will be located at $(1.5)^2$ or 2.25 millimeters to the right of the 0-mark. 2.5 will be located at $(2.5)^2$ or 6.25 millimeters from 0; and so on.

With two such scales, you can use them like a slide rule.

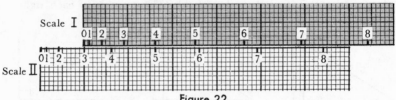

Figure 22

The settings in Figure 22 show the solution for the hypotenuse of a right triangle having sides of 3 units and 4 units. The answer, 5, is obtained by adding the scale lengths 3 and 4. Set the 0 reading of Scale I over the 3-mark of Scale II. Find the 4-mark on Scale I and under it, on Scale II, read the answer, 5.

The same settings of the slide rule show how to find the square root of $5^2 - 4^2$ by subtracting length 4 from length 5. The 4-mark of Scale I is lined up with the 5-mark on Scale II. The answer, 3, is found under the 0 mark of Scale I.

Build yourself a slide rule and find the solutions to Exercises 1-7 of Exercise Set 9.

Exploring Two Special Right Triangles

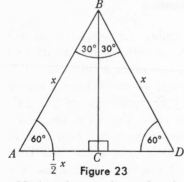

Figure 23

Two triangles that a draftsman uses as standard equipment are the 30°-60° and the 45° right triangles.

Consider first the 30°-60° right triangle. Place two 30°-60° right triangles in the position shown in Figure 23. The corresponding sides of these triangles are equal.

Notice that each angle of triangle ADB contains 60°. Thus triangle ADB is equilateral, and $AB = BD = AD$. Since one of our initial conditions was that $AC = CD$, then $AC = \frac{1}{2}AD = \frac{1}{2}AB$. In other words, the length of the shortest leg of a 30°-60° right triangle is one-half the length of the hypotenuse.

What is the relationship between BC and AB in the above drawing? Here is how you can find the answer:

Let $AB = x$. Then $AC = \frac{1}{2}x$.

By the Pythagorean theorem,

$$BC = \sqrt{(AB)^2 - (AC)^2}$$

$$= \sqrt{x^2 - \left(\frac{1}{2}x\right)^2}$$

$$= \sqrt{x^2 - \frac{1}{4}x^2}$$

$$= \sqrt{\frac{3}{4}x^2}$$

$$= \frac{\sqrt{3}}{2}x$$

$$= \frac{1.732}{2}x$$

$$= .866x.$$

The 30°-60° right triangle is used so frequently in engineering work that the relationships among the sides are memorized, and either one of the following figures is used to aid the memory:

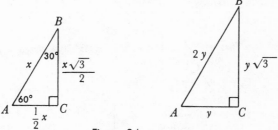

Figure 24

Thus, if $AB = 60$ feet, then the other sides can be computed immediately:

$$AC = \frac{1}{2}(60 \text{ feet}) = 30 \text{ feet}$$

$$BC = \frac{\sqrt{3}}{2}(60 \text{ feet}) = .866(60 \text{ feet}) = 52 \text{ feet (to the nearest foot)}.$$

In somewhat the same way, you can find the relationships between the sides of the 45° right triangle.

127

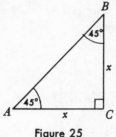

Figure 25

In this case, the two legs, AC and BC, are equal. Let each be x. Then by the Pythagorean theorem,

$$AB = \sqrt{x^2 + x^2}$$
$$= \sqrt{2x^2}$$
$$= \sqrt{2}(x)$$
$$= 1.414x.$$

The figure that helps an engineer to remember these relationships is shown in Figure 26.

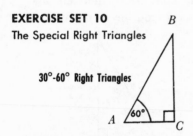

Figure 26

Suppose you had a square whose sides were 90 feet. You can then find the diagonal of the square immediately by applying the formula:

Diagonal $= \sqrt{2}(90 \text{ feet}) = 1.414(90 \text{ feet})$
$= 127$ feet (to the nearest foot).

EXERCISE SET 10
The Special Right Triangles

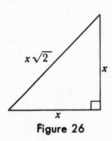

30°-60° Right Triangles

45° Right Triangles

1. *a.* GIVEN: $AB = 30$ feet
FIND: AC and BC

2. *a.* GIVEN: $AC = 30$ feet
FIND: AB and BC

3. *a.* GIVEN: $BC = 30$ feet
FIND: AB and AC

1. *b.* GIVEN: $AB = 30$ feet
FIND: AC and BC

2. *b.* GIVEN: $AC = 30$ feet
FIND: AB and BC

3. *b.* GIVEN: $BC = 30$ feet
FIND: AB and AC

4. GIVEN: $BC = 16$ feet and angles as shown.
FIND: AC, CD, BD, and AB.

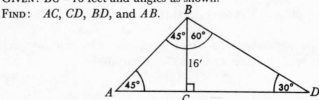

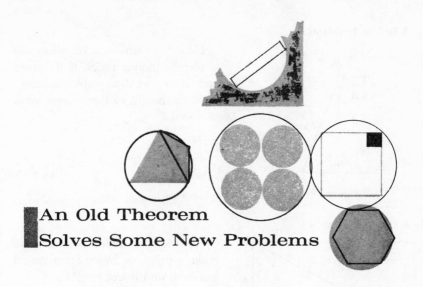

An Old Theorem
Solves Some New Problems

There are many interesting problems that can be solved by finding the length of one side of a right triangle in terms of the length of the other two sides. A number of these applications of the Pythagorean theorem are illustrated by the following problems.

Measuring a Lake

To find the distance between points A and B on opposite sides of a small lake, a man set a stake at point C so that angle B is a right angle. By measuring, he found AC to be 160 feet and BC to be 128 feet. How far is it between A and B?

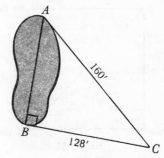

Figure 27

How Much Cable?

How many feet of cable will be necessary to reach from point A, 27 feet up on a pole, to the point C on the ground 36 feet from the base B of the pole?

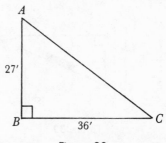

Figure 28

129

A Ladder Problem

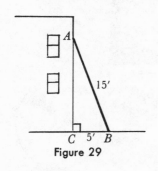

Figure 29

How high up on a building will a 15-foot ladder reach if its lower end is 5 feet from the building? Find the height to the nearest tenth of a foot.

A Gate Problem

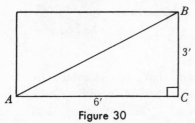

Figure 30

How long a board is needed to make a diagonal brace for a gate 3 feet high and 6 feet long?

Diamond Dimensions

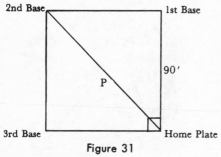

Figure 31

A baseball field is a square, each side of which is 90 feet long. What is the distance from home plate to second base?

The pitcher's box is 60 feet from home plate. How far is it from the pitcher's box to each of the other bases?

Crossing the River

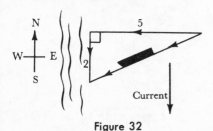

Figure 32

A boat is steered due west across a river which has a current of 2 miles per hour south. The boat moves 5 miles per hour with respect to the water, but the current causes it to move several degrees south of west. What speed does the boat have with respect to the land?

Figure 33

A traveling crane in a factory can move heavy machinery to any part of the floor. The crane extends across the width of the building and rolls on tracks which run along the sidewalls just under the eaves. A machine can be moved along the crane across the building with a velocity of v_x, and at the same time the crane can move the length of the building with a velocity of v_y. The two velocities are at right angles to each other.

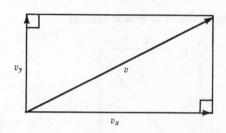

Figure 34

The velocity v of the machine with respect to the floor is called the *vector sum* of these two and is shown as the diagonal of the rectangle. Then the magnitude of the velocity v is

$$v = \sqrt{v_x{}^2 + v_y{}^2}.$$

Find v if $v_x = 2$ feet per second and $v_y = 4$ feet per second.

A Problem in Forces

Two forces, F_1 and F_2, at right angles to each other, are applied on an object. The effect of these two forces is the same as if a single force F were applied to the object. F is called the *resultant* of the two forces. Find F when $F_1 = 7$ pounds and $F_2 = 6$ pounds.

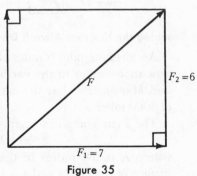

Figure 35

131

A Rope Problem

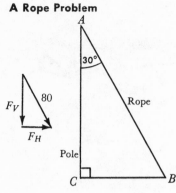

Figure 36

A rope is attached to the top of a pole AC at point A. A man pulls on the rope at point B so that the rope makes a 30° angle with the pole. If the pull along the rope is 80 pounds, what are the horizontal, F_H, and vertical, F_V, components of the force?

Mathematics of a Wall Bracket

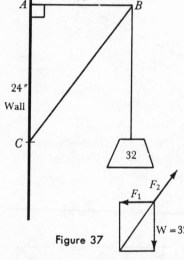

Figure 37

In a simple wall bracket, where $AB = 18$ inches, $AC = 24$ inches, and a weight of 32 pounds is hung at B, the weight of 32 pounds tends to pull the fastener out at A with a force of F_1. There is also a thrust, F_2, along BC. The forces, F_1, F_2, and W are in balance and are directly proportional to the corresponding dimensions, AB, BC, and AC. Find BC by the Pythagorean theorem and then determine the forces from the following ratios and proportions:

$$\frac{F_1}{18} = \frac{F_2}{BC} = \frac{32}{24}.$$

Scanning the Horizon: Aircraft Division

An airplane pilot 5 miles above the surface of the earth can see how many miles to the earth's horizon? Assume perfect visibility and also assume that the earth is a perfect sphere with a radius of 4000 miles.

The Pythagorean theorem says

$$x^2 + R^2 = (h + R)^2,$$

where R is the radius of the earth, h is the distance above the surface of the earth, and x is the distance from the point above the

earth to the horizon. By expanding $(h+R)^2$ and combining like terms, this formula can be changed to
$$x^2 = h^2 + 2hR.$$

Since h^2 is very small in comparison to $2hR$, you will obtain a good approximation to x by dropping h^2. This reduces the formula to the very simple expression:
$$x^2 = 2hR$$
$$\text{or } x = \sqrt{2hR}.$$

Now you can compute the value of x for $h = 5$ miles and $R = 4000$ miles. It is important that h and R be measured in the same units.

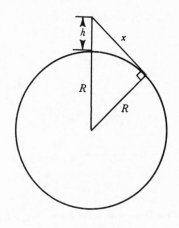

Figure 38

Scanning the Horizon: Down-to-Earth Division

A man whose eyes are 6 feet above sea level looks out to sea on a clear day. How far can he see to the horizon? The same formula can be used that is given in the previous problem, namely, $x = \sqrt{2hR}$. But what happens to this formula when h is expressed in feet? We can express the height in miles by dividing h feet by 5,280 feet per mile. Substituting this in the formula, we have:

$$x = \sqrt{2 \times \frac{h}{5280} \times 4000},$$
$$= \left(\sqrt{\frac{8000}{5280}}\right)(\sqrt{h}),$$
$$= (\sqrt{1.5151})(\sqrt{h}),$$
$$= 1.23\sqrt{h},$$

where h is expressed in feet and x in miles.

Now use this formula to find x when $h = 6$ feet.

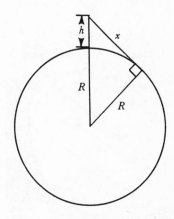

Figure 39

The Stick and the Trunk

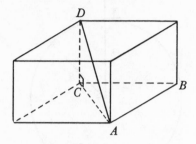

Figure 40

The Diameter of a Pipe

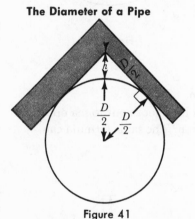

Figure 41

A Machinist's Brain Teaser

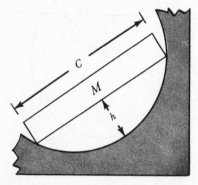

Figure 42

What is the longest stick, to the nearest inch, that can be placed inside a trunk 24 inches wide, 30 inches long, and 18 inches high. *Hint:* Put the following two equations together:

$$\overline{AC}^2 = \overline{AB}^2 + \overline{BC}^2$$
$$\overline{AD}^2 = \overline{AC}^2 + \overline{DC}^2$$

to make one equation

$$\overline{AD}^2 = \overline{AB}^2 + \overline{BC}^2 + \overline{DC}^2.$$

You can find the diameter, D, of a pipe by placing the pipe in the corner of a carpenter's square, measuring the distance h from the inside corner to the pipe, and then multiplying this length by 4.828; that is,

$$D = 4.828h.$$

Prove that this formula is correct, then test it out on several different sizes of pipe.

A machinist has to find the diameter of the circular arc of a hollow portion of an object. He places a convenient straightedge of known length C as shown in the figure. He bisects C at M and measures the height h of this chord. He can then compute the diameter D of the circular arc from the formula

$$D = h + \frac{C^2}{4h}.$$

Prove that this formula is correct.

Hint: Draw a full circle with center at O and a radius $OB = \dfrac{D}{2}$. Since $AB = C$ and M is the midpoint of AB, then $MB = \dfrac{C}{2}$. Since ON is also a radius, then $ON = \dfrac{D}{2}$. MN is given as h. Therefore, $OM = \dfrac{D}{2} - h$.

You now have a right triangle OMB whose three sides are expressed in terms of D, h, and C. Now use the Pythagorean theorem to find D in terms of C and h.

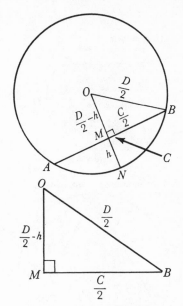

Figure 43

An interesting project would be to draw a series of circles of different diameters, measure a chord AB and the corresponding distance MN, and compute the diameter from these two measured lengths. Check the computed value against the actual diameter of the circle. Try different lengths of chords in the same circle. Notice that as the chord changes, so does the value of h, but the diameter should still come out the same. Of course, since measurements may have some error, you can expect to have some differences, but they should be slight if you do careful work.

A Problem with Gears

A machinist wants to place four equal gears inside a circular casing so that they are all tangent as shown. If the outer circle has a diameter of D, and all four inner circles have a diameter of d, then prove that

$$D = 2.414d.$$

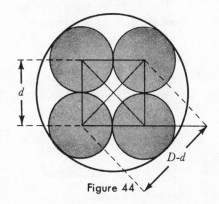

Figure 44

135

Milling a Bar

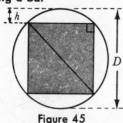

Figure 45

A square shaft is milled (cut) from a round bar. Find the depth of the cut h in terms of the diameter of the bar. In other words, prove that

$$h = .146D.$$

A Circle, a Hexagon, and a Triangle

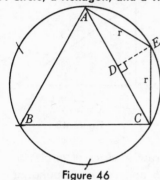

Figure 46

Construct a regular hexagon inscribed in a circle by making the sides all equal to the radius of the circle. Join every other vertex of the hexagon to form the triangle ABC. Prove that $AC = 1.732r$.

Hint: Each angle of a regular hexagon contains 120°. Let perpendicular DE bisect angle AEC.

A Hexagon Fact

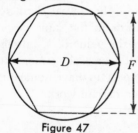

Figure 47

Prove that the distance across the flat, F, of a regular hexagon is .866 times the dimension D.

From a Square to an Octagon

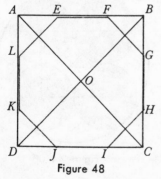

Figure 48

Draw a square. Connect opposite corners AC and BD. Label their point of intersection O. Using AO as radius, and A as center, draw arcs intersecting AB at F and AD at K. Similarly, locate points E, G, H, I, J, and L, by using centers at B, C, and D. Draw octagon $EFGHIJKL$ and prove that all sides of the octagon are equal.

The Maltese Cross

The Maltese Cross is a figure made of five equal squares arranged as shown in Figure 49. Cut it into the fewest possible pieces which can be arranged to form a single square.

Hint: The area is 5 square inches. The side of the square must therefore be $\sqrt{5}$ inches.

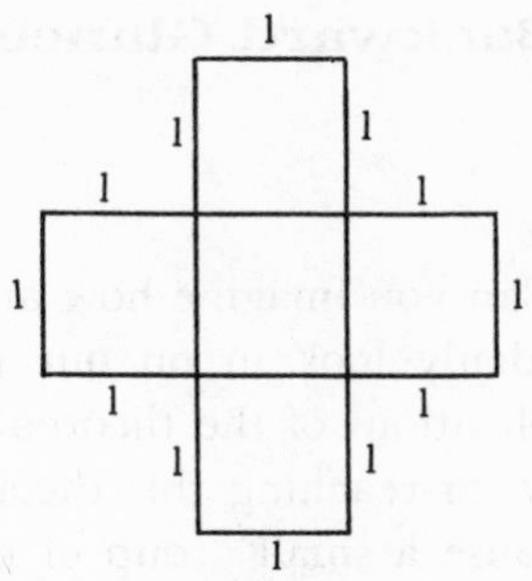

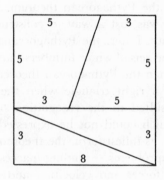

Figure 49

The Mystery of the Increased Area

A square 8×8 is cut as shown in Figure 50.

Figure 50

The pieces are rearranged to form a triangle as shown in Figure 51. The area of the square is 64. The area of the triangle seems to be 65. Try to explain this situation.

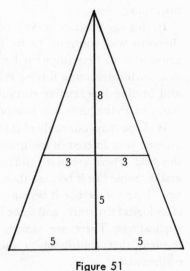

Figure 51

A Backward Glance and a Look to the Future

Can you imagine how amazed Pythagoras might be if he could suddenly look in on our modern space age and see the many applications of the theorem named after him? Little did he know how far-reaching this theorem might be. And to think, it began because a small group of men, called Pythagoreans, had enough curiosity to look for interesting relationships with whole numbers.

Discovery techniques played a major role in the developing of the Pythagorean theorem. At first, it wasn't really a theorem. It was just a way to construct a right angle to locate property lines. Later, the Pythagoreans formulated a fascinating puzzle to find sets of whole numbers that satisfied the relationship $a^2 + b^2 = c^2$. Then the Pythagorean theorem became useful in finding any side of a right triangle when the other two sides were given. This resulted in the recognition of a new kind of number like $\sqrt{2}$ which could not be expressed as a ratio of two whole numbers. In its fullest form, the theorem became useful in solving a variety of problems — finding inaccessible distances, finding the resultant of forces and velocities, and even solving puzzle-type problems. Yes, the Pythagorean theorem became extremely useful as well as intriguing.

In this age of space travel, you can be sure that the Pythagorean theorem will continue to be a useful tool. In fact, it would be impossible to be without it because space travel must of necessity deal with interacting forces, velocities, accelerations, decelerations, and finding inaccessible distances. Thus you can be sure that the name of Pythagoras will live on forever.

We hope that your study of the material in this section has sufficiently aroused your interest in mathematics to cause you to want to go on in this field. There are many different reasons why people enjoy mathematics. Some like it because there is always a real challenge to discover new things; some like it because of its simplicity; some like it because of its logical structure; and some like it because it has so many practical applications. There are elements in the study of the Pythagorean theorem that contribute to each one of these reasons for liking mathematics.

PART IV

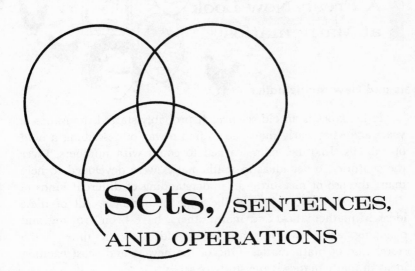

Sets, SENTENCES, AND OPERATIONS

A Fresh New Look at Mathematics

Sets and New Mathematics

Mathematics is an old science. It probably started thousands of years ago when early man had to find a way of describing a herd of animals. In time, man learned to count with numbers. Over the centuries, other ideas in mathematics were developed to help man: the use of measures, an understanding of different kinds of shapes, and a knowledge of the logic of numbers. All of these ideas in mathematics, and many others, have been growing and changing, and new ideas are being added all the time to the storehouse of mathematics. One of the newest and most exciting ideas in mathematics is the study of sets.

What do we mean by a set? A *set* is simply a group or a collection of things. This idea of grouping things is not new to you. You already know the meaning of these expressions: "a set of books," "a collection of stamps," "a chain of stores," "a set of exercises." Objects are usually grouped because of some common feature. The set of books may all be encyclopedias; the stamps were perhaps issued by the same country; the stores in the chain are probably owned by the same company; the exercises may cover the same kind of material. We sometimes use other words to talk about collections or sets, such as *flock, herd, gang, team, company, group, club, class,* or *family*. You are even a member of groups or sets like these: your family, your class, or your scout troop.

The mathematician calls all of these collections by the term *set*. A set may be a collection of objects, numbers, persons, figures, or ideas. The objects or numbers making up a set belong to the set.

These items which belong to the set are called *members* or *elements* of a set.

The mathematical theory of sets began during the late years of the nineteenth century. A German mathematician, Georg Cantor, tackled a difficult mathematical problem concerned with the study of endless quantities. It involved such questions as: "How many whole numbers are there?" "How many points are there on a circle?" "How many instants of time elapse in one hour?" "Are there more numbers between 1 and 2 than there are points on a line?" Cantor solved his problem, and his work marked the beginning of the set idea in mathematics.

As sometimes happens with a new idea, Cantor's findings were not accepted at first. He was ridiculed for many years, but by 1920, his new way of thinking had gained some recognition from mathematicians. And now, the theory of sets has been applied to many fields of mathematics from algebra to probability.

The science of sets tells us how to combine sets and how to compare sets to determine relationships. Solving equations, drawing graphs, studying chance or probability, describing geometric figures become simpler by using the language and ideas of sets. Particularly exciting is the use of sets in solving problems, puzzles, and mysteries with electronic computers.

One of the first things you need to learn in the study of sets is to identify the members of a set. The exercise below will help you select members of sets.

EXERCISE SET 1
Discovering Sets

1. Write the members of each of these sets:
    a. the set of vowels.
    b. the set of odd numbers less than 10.
    c. the set of months beginning with the letter M.
    d. the set of United States senators from your state.

2. What are the members of the set of whole numbers less than 20 which are prime numbers? Prime numbers are whole numbers, except 1, evenly divisible only by themselves and 1.

3. Which of the following are members of the set of General Motors cars: Ford, Chevrolet, Rambler, Buick, Dodge, Pontiac?

4. a. How many members does the set of letters of our alphabet have?
    b. How many members does the set of vowels of our alphabet have?

5. a. Write the members of the set of numbers less than 99 which are divisible by 9, such as 9, 18, 27, and so on.
    b. Write the set of numbers less than 99 in which the sum of the digits is 9.
    c. Do the sets in a and b above have the same members?

The Sign Language of Sets

Suppose that we consider the set of odd numbers from 1 through 9. This set has the members 1, 3, 5, 7, 9. Mathematicians like to translate statements like this into symbols. The notation $\{1, 3, 5, 7, 9\}$ is another way of writing the statement "the set of odd numbers from 1 through 9." The symbols [ ] or { } are used to group the members of a set. In this pamphlet, the second symbol, { }, will be used. Usually we use a capital letter to designate a set, like this: $A = \{1, 3, 5, 7, 9\}$. The notation $\{1, 3, 5, 7, 9\}$ is called a *tabulation* of the members of the set.

It is important that we be able to translate a statement into a set tabulation, but we must also be able to change a tabulation into a statement. For example, the tabulation $\{a, e, i, o, u\}$ can be translated as "the set of vowels of our alphabet." When we use a capital letter to designate the set, such as $X = \{a, e, i, o, u\}$, we read this as "X is the set whose members are $a, e, i, o,$ and u" or "X is the set of vowels of our alphabet." The way we describe a set should tell us what items belong to the set and what items are not members of the set.

Set Members

1. List or tabulate the members of the following sets:
   a. the names of the states of the United States that begin with the letter M.
   b. the set of symbols used in Roman numerals.
   c. the set of presidents of the United States whose last names begin with J.
   d. the set of the squares of all whole numbers between 0 and 10.

2. Describe the following sets in words:
   a. $K = \{ 5, 10, 15, 20, 25 \}$
   b. $M = \{ 3, 6, 9, 12, 15, 18 \}$
   c. $R = \{ a, b, c, d, e \}$
   d. $Y = \{ \text{Tuesday, Thursday} \}$

3. Answer these questions about members of a set:
   a. Is a silver dollar a member of the set of United States coins?
   b. Is a whale a member of the set of all fish?
   c. Is meat a member of the set of foods containing protein?
   d. Is 64 a member of the set of squares of whole numbers?

Comparing Sets

In arithmetic, we often compare numbers to find out which one is the larger and how much larger. When we write equations or mathematical sentences like $2 + 3 = 5$ or $2x + 7 = 9$, we are interested in equivalent quantities. In a similar way, we solve problems with sets by comparing one set with another. In the study of chance or probability we need to know when sets are equal or how many members a set has. When solving problems by reasoning, we need to know if the members of one set are also the members of another set. In scientific research, the scientist needs to know how the members of one set of data are related to the members of another set.

One way to compare sets is to compare the *members* of one set with the *members* of another set. The sets $\{ a, b, c \}$ and $\{ x, y, z \}$ are different because they have different members. The sets $P = \{ a, b, c \}$ and $Q = \{ c, a, b \}$ are *equal* because they have the same members. We write this $P = Q$. It doesn't make any difference whether the members are in the same order or not. Any two sets with the same members are equal.

It is easy to recognize equal sets when the members are listed. But it is not always easy to decide about equality when the sets are described in words.

EXERCISE SET 3
Finding Equal Sets

Compare these sets to find out if they are equal:

1. $\{\,5, 10, 15, 20, 25\,\}$ and $\{\,5, 15, 25, 10, 20\,\}$
2. $R = \{\,a, h, m, t\,\}$ and $T = \{\,m, a, t, h\,\}$
3. $\{\,1, 5, 7, 9\,\}$ and the set of odd numbers less than 10.
4. the set of students in your class with an "A" average in mathematics and the set of boys in your class.

One-to-one Matching

Long ago shepherds kept track of their sheep by matching a pebble with each sheep. The number of pebbles represented the number of sheep in the flock. The matching of a pebble with a sheep gave a one-to-one relationship between the two. This was really a comparison between two sets: a set of sheep and a set of pebbles.

A comparison of any two sets may be made by matching members of one set with members of another set. The sets $A = \{\,a, b, c\,\}$ and $B = \{\,x, y, z\,\}$ are not equal. The members of A are not the same as the members of B, but each set has the same number of members. Every member of set A can be matched with a member of set B like this:

$$\{\, a, \quad b, \quad c \,\}$$
$$\updownarrow \quad \updownarrow \quad \updownarrow$$
$$\{\, x, \quad y, \quad z \,\}$$

144

This matching or pairing of members gives us what is called a *one-to-one correspondence.*

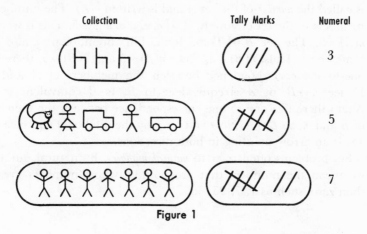

Collection	Tally Marks	Numeral

Figure 1

The collections and tally marks in Figure 1 show how a one-to-one correspondence between tally marks and sets of objects is related to our symbols for numbers.

A numeral is a symbol for a number and is often called "a name for a number." Sometimes we say a numeral describes a set. The numeral "5" describes all sets which have a one-to-one correspondence to five tally marks or five units. Frequently the word "number" is used to refer either to a number or to a name for it.

Whenever one set has a one-to-one correspondence to another set, the sets are said to be *equivalent.* For example, the sets $A = \{$ Tom, Bill, Mark $\}$ and $B = \{$ bicycle, skates, football $\}$ are equivalent. They are not equal because the members are different. But by matching or pairing members we get a one-to-one correspondence, like this:

$$A = \{\text{Tom,} \quad \text{Bill,} \quad \text{Mark} \}$$
$$B = \{\text{bicycle,} \quad \text{skates,} \quad \text{football} \}$$

We write this equivalence this way: $A \leftrightarrow B$. We read this "set A is equivalent to set B." Remember that equivalent and equal have different meanings.

Sometimes we also talk about a one-to-*two* correspondence. An example of one-to-two correspondence would be the set of people in a room and the set of hands of these people.

An easy way to find whether sets are equivalent or not is to count the members of each set. The number of members of set A is called the *number* of the set A and is written $n(A)$. The number of members in the set of vowels, $A = \{a, e, i, o, u\}$, is 5. This is written $n(A) = 5$. The set $B = \{$Tom, John, Bill, Brent, Gary$\}$ also has 5 members. Thus, $n(B) = 5$. Since $n(A) = 5$ and $n(B) = 5$, there is a one-to-one correspondence between the members of A and B. Hence $A \leftrightarrow B$, or A is equivalent to B. Is B equivalent to A? Again there is a one-to-one correspondence between the members of B and A, so $B \leftrightarrow A$. Now you see why the symbol of equivalence, $\leftrightarrow$, is an arrow pointing in both directions.

Set problems often refer to *natural numbers*. By natural numbers we mean the numbers that are positive whole numbers greater than zero, such as 1, 7, 95, 368.

EXERCISE SET 4
Making Set Comparisons

1. Which of these sets are equal?
   a. $\{a, p, r, t\}$ and $\{r, a, p, t\}$
   b. $\{1, 2, 3, 4\}$ and $\{2, 4, 6, 8\}$
   c. the set of prime numbers less than 10 and $\{1, 2, 3, 5, 7\}$
   d. $\{$California, Oregon, Alaska, Washington, Hawaii$\}$ and the set of states with borders on the Pacific Ocean.

2. Which of these pairs of sets have a one-to-one correspondence?
   a. $\{a, b, c, d, e\}$ and $\{3, 7, 4, 8, 5\}$
   b. the set of odd numbers below 20 and the set of prime numbers below 20.
   c. $\{$Jim, Dave, Mike$\}$ and $\{$Mary, Cathy, Karen, Molly$\}$
   d. the set of even natural numbers less than 20 and the set of odd natural numbers less than 20.

3. What is the number of members of each of these sets?
   a. $A = \{$shoe, stocking, boot, rubbers, slipper$\}$
   b. the set of states of the United States.
   c. the set of United States senators.
   d. the set of natural numbers that divide 24 without a remainder.

4. Which of these sets are equivalent?
   a. $\{1, 2, 3, 4, 5, 6\}$ and $\{7, 14, 21, 28, 35, 42\}$
   b. $\{5, a, \not{X}, \triangle, \sim, \xi\}$ and $\{J, R, P, B, C, M\}$

c. the set of states of the United States, and the set of United States senators.

d. $\{\frac{1}{2}, \frac{1}{3}, \frac{1}{4}, \frac{1}{5}, \frac{1}{6}\}$ and $\{\frac{3}{4}, \frac{3}{5}, \frac{3}{7}, \frac{3}{8}, \frac{3}{10}\}$

5. Find the number of members of these sets:

   a. n { the set of letters in our alphabet }

   b. n { the set of even natural numbers less than 50 }

   c. n { the set of prime numbers between 30 and 40 }

   d. n { the set of number symbols in the decimal number system }

Finite, Infinite, and Empty Sets

For some sets, the number of members is so large that it would be very tiresome to list them all; for example, the set of people in your town, or the set of natural numbers less than 1000. We use dots to show that we have left some numbers out:

$$\{1, 2, 3, 4, 5, \ldots 999, 1000\}.$$

Sometimes there is no end to the number of members in a set; for example, the set of natural numbers. We say this set has an *infinite* number of members. We write this set this way: $N = \{1, 2, 3, 4, 5, \ldots\}$. We cannot count the number of members of this set. Another way of saying this is that the set of natural numbers is an infinite set. When a set is not infinite, we can count the members in the set with a number and we call it a *finite* set. The number of members in an infinite set is always greater than any number we can count. Some examples of infinite sets are: the set of points on a line, the set of minutes in the future, the number of prime numbers, the set of even numbers.

Let's compare the members of two infinite sets. Let $N = \{1, 2, 3, 4, 5, \ldots\}$ be the set of all natural numbers and $E = \{2, 4, 6, 8, 10 \ldots\}$ be the set of even natural numbers. What happens when we match or pair members of these sets?

$$N = \{1, 2, 3, 4, 5, \ldots\}$$

$$E = \{2, 4, 6, 8, 10, \ldots\}$$

We see that every natural number can be matched with an even number. This seems to indicate that there are as many even numbers as there are natural numbers! This is expected, since an operation (multiplication by 2) changes the members of one set to the other.

Is it possible for the collection of even numbers, which is contained in the set of natural numbers, to have as many members as the set of natural numbers? For many years, this problem troubled the greatest mathematicians of the world. But it was solved when Georg Cantor, the German mathematician mentioned earlier in this booklet, proposed his ideas on infinite collections that became the basis for set theory. He showed that the distinguishing feature of an infinite set is that it can be placed in a one-to-one correspondence with a part of itself. Although the conclusion that there are as many even numbers as there are natural numbers is difficult to accept, set ideas show it to be logical.

In contrast to infinite sets, there are also sets with no members. For example, what is the set of even numbers that are divisors of 13? This set is called an *empty set* or *null set*.

The empty set is similar to zero in our number system. It is usually shown by the symbol ϕ, which is a zero with a line through it. The empty set can also be shown like this: { }. Other examples of the empty set are: the set of living persons more than 200 years old, the set of rectangles with five sides, the set of even prime numbers between 20 and 30.

EXERCISE SET 5
Infinite Sets and Empty Sets

1. Which of the following sets are infinite?
   a. the grains of sand on all the beaches of the world.
   b. the population of the world.
   c. the atoms in our world.
   d. the natural numbers evenly divisible by 13.

2. Which of the following sets have a one-to-one correspondence?
   a. { 1, 2, 3, 4, . . . } and { 5, 10, 15, 20, . . . }
   b. the set of natural numbers and the set of squares of natural numbers.
   c. the set of words in the English language and { 1, 2, 3, 4, . . . }
   d. the set of multiples of 5 and the set of multiples of 1,000,000.

3. Which of the following sets are empty sets?
   a. the set of foreign-born presidents.
   b. the set of men 15 feet tall.

c. the set of even prime numbers.

d. the set of squares of odd numbers that are even.

4. Name some sets which have no members.

Sets within Sets

Jill was talking about last Friday's party for the ninth grade. She said, "Jane and Ruth had beautiful blue dresses. James and Chuck came late. The quartet of 4D's — Dill, Dan, Dick, and Dave — won the talent contest."

Jill was talking about different groups who were members of the set of ninth graders. These groups, such as Jane and Ruth, are called *subsets* of the set of ninth graders. Jane and Ruth are members of the set of ninth graders and also members of the subset of ninth-grade girls with beautiful blue dresses. Likewise, James and Chuck are members of a subset, the subset of tardy boys. Similarly, the 4D's are a subset of the ninth graders participating in the talent contest.

In the set $K = \{\, a, b, c, d, e \,\}$, we may wish to consider the vowels a and e. The letters a and e may be called the set of vowels of the set K. We say that a and e are members of a subset of K. A subset is part of a set. Each member of the subset is also a member of the whole set.

A Homecoming Committee is made up of this set: $\{\,$ John, Sue, Carla, Ralph, David $\,\}$. The set of boys on the committee, $\{\,$ John, Ralph, David $\,\}$, is a subset of the Homecoming Committee set. Other subsets of this committee could be $\{\,$ Sue, Carla $\,\}$ and $\{\,$ John, Sue, David $\,\}$.

Sometimes we need to know how many subsets can be formed for a given set. This is like asking how many committees we could have made up of these people: $\{\,$ John, Ralph, David $\,\}$. Here they are:

$A = \{\,$ John, Ralph, David $\,\}$ $E = \{\,$ John $\,\}$

$B = \{\,$ John, Ralph $\,\}$ $F = \{\,$ Ralph $\,\}$

$C = \{\,$ John, David $\,\}$ $G = \{\,$ David $\,\}$

$D = \{\,$ Ralph, David $\,\}$ $H = \{\quad\}$

149

You may think it strange to think of a committee of one or a committee of none. In sets, this is possible as long as we don't say how many members each committee should have. Notice also that we think of the original set as a subset of itself. The members of every subset are members of the original committee. So when we ask how many subsets can be formed, we include the empty set and the original set.

Sometimes the set of subsets which can be formed from a set is called the *power* set of the original set. This idea has many applications in the mathematics of chance or probability, where the number of subsets tells us the number of ways different events may occur.

EXERCISE SET 6
Working with Subsets

1. If $A = \{ 1, 3, 5, 7, 9 \}$, answer the following questions:
   a. What is the subset of prime numbers in set A?
   b. What is the subset of numbers evenly divisible by 3 in set A?
   c. What is the subset of numbers evenly divisible by 1 in set A?
   d. What is the subset of numbers evenly divisible by 2 in set A?

2. Make up one subset for each of the following sets:
   a. $\{ w, x, y, z \}$
   b. $\{$ Ford, Chevrolet, Plymouth, Rambler $\}$
   c. the set of holidays in a year.
   d. the set of four-sided geometric figures.

3. List all the possible subsets for each of the following sets:
   a. $\{ a, b, c \}$ c. $\{$ Peg, Sue, Kay $\}$
   b. $\{ 7, 11 \}$ d. $\{ p, g, r, s \}$

4. Copy and complete this table, showing the number of subsets for each set:

Set	Number of Members of the Original Set	Number of Subsets
$\{ a \}$	1	
$\{ a, b \}$	2	
$\{ a, b, c \}$	3	
$\{ a, b, c, d \}$	4	

5. a. How many subsets do you think there will be if the original set in Example 4 has 5 elements?
   b. Can you write a formula for the number of subsets, P, if the number of members of the original set is n?

The Universal Set

Whenever we talk about sets, we have in mind a collection of things which have some common characteristic. For example, we may be talking about even numbers, basketball players who are 7 feet tall, or triangles. In these sets, the common characteristic of each is evident: the numbers are divisible by 2; the people are all 7 feet tall; and the triangles are geometric figures with three sides. However, the members of each of these sets can be considered subsets of a much larger set, called the *universe* or *universal set*. For example, the even numbers are a subset of the universal set of all numbers; the tall basketball players are a subset of the universal set of all basketball players; and the triangles are a subset of the universal set of all geometric figures.

Sometimes it is not clear what the universal set is unless we describe it specifically. Suppose the members of a set of cars are all Chevrolets. The universal set may be all General Motors cars, all American cars, all low-priced cars, or all cars in the world.

The universal set is designated by the capital letter U. If, for a set of numbers, no other description is given for the universal set, it is usually interpreted as "the set of all real numbers" or $U = \{$ all real numbers $\}$. This set includes all positive and negative whole numbers, fractions, and irrationals (numbers that cannot be expressed as the ratio of two whole numbers, such as $\sqrt{2}$).

EXERCISE SET 7
Universal Set Problems

1. What is a universal set for these sets?
   a. the set of students at Central High School with "A" grades.
   b. the set of Ford cars.
   c. the set of cashmere sweaters.
   d. the set of whole numbers between 10 and 20.
2. Name a subset of each of the following universal sets:
   a. the set of students in your school.
   b. the set of whole numbers greater than 3 and less than 30.
   c. the set of four-sided figures.
   d. the set of books in your library.

Problems Solved
by Sets and Pictures

Set Diagrams

"One picture is worth a thousand words" is a very old saying. When solving mathematical problems, we often have to find relationships between sets of values, objects, or events, and a good way to see how sets are related is to represent them with drawings.

Mathematicians usually use a rectangle to represent a universal set. All the members of the universal set are considered as points on or inside the rectangle. For example, U = the set of all people in Crow Valley, is diagramed in Figure 2.

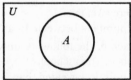

Figure 2

A subset of the universe is usually represented by a circle. Figure 3 shows how to represent A, the set of all men in Crow Valley.

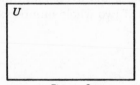

Figure 3

A subset of set *A* is represented by another circle. Let *B* = the set of men of Crow Valley who are older than 50 years of age. Figure 4 pictures the relationship between the sets.

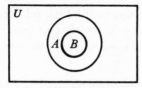

Figure 4

These set diagrams are called *Venn diagrams.*

If you stop to think about the ways any two circles can be related, you will know how any two sets can be related. They can be related in these ways:

1. Circles can be separate, like those in Figure 5.

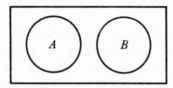

Figure 5

Then we say that *A* misses *B*, or *A* does not intersect *B*. The sets represented are completely separate and independent, with no members in common. Set *A* is said to be *disjoint* from set *B*.

An example of disjoint sets is:

X = the set of boys in our class

Y = the set of girls in our class.

Another example is:

A = the set of all odd numbers

B = the set of all even numbers.

2. Circles can meet, like those in Figure 6.

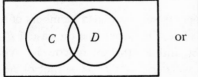 or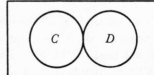

Figure 6

Then we say set C meets or *intersects* set D. This happens when some of the members of set C are also members of set D.

An example of intersecting sets is:

C = the set of boys in the tenth grade

D = the set of boys in the high school science club.

If C and D intersect, we know that at least one tenth-grade boy is also a member of the science club. We assume that the club has members from other classes.

3. One circle can completely coincide with another, as in Figure 7.

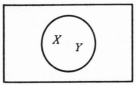

Figure 7

When circles coincide, the sets represented have exactly the same members. We know that sets with exactly the same members are said to be identical or equal. We write this relationship as $X = Y$.

If X = the set of even natural numbers less than 20 and Y = the set of natural numbers less than 20 which are divisible by 2, then

$$X = \{ 2, 4, 6, 8, 10, 12, 14, 16, 18 \},$$
$$\text{and } Y = \{ 2, 4, 6, 8, 10, 12, 14, 16, 18 \},$$
$$\text{or } X = Y.$$

4. One circle can be inside the other, as in Figure 8.

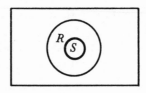

Figure 8

This means that all members of set S are also members of set R. In other words, S is a subset of R. This would be the case if R = the set of letters of the alphabet and S = the set of vowels. When S has fewer members than R, we say S is a *proper* subset of R. This relationship is written with symbols this way: $S \subset R$. The expression $S \subset R$ is read "S is included in R" or "S is a subset of R."

When more than two sets are related, we can still show the relationships with drawings. Let's see how this works with these three sets:

Let A = the set of all boys in your state,

B = the set of all boys in your school,

C = the set of all boys in your mathematics class.

These sets are related like the circles in Figure 9.

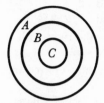

Figure 9

Here is another kind of relationship:

Let X = the set of girls in your school,

Y = the set of girls named Mary,

Z = the set of girls who are 15 years old.

These sets would probably be related as shown in Figure 10.

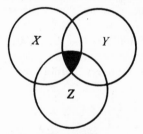

Figure 10

This diagram says that some of the girls in your school are named Mary, some of the girls in your school are 15 years old, and some of the girls in your school both are named Mary and are 15 years old. The solid area represents this last group.

EXERCISE SET 8
Set Diagrams

1. Make Venn diagrams to represent the relationships of the given pairs of sets:

a. the set of General Motors cars; the set of Chevrolets.

b. the set of Chevrolets; the set of Fords.

c. the set of Plymouths; the set of 1960 model automobiles.

d. the set of players on your school's football team; the set of players on your school's basketball team.

2. Show the relationships between the following sets with Venn diagrams:

a. $\{\,2, 4, 6, 8\,\}$ and $\{\,1, 3, 5, 7, 9\,\}$

b. $\{\,1, 2, 3, 4, 5\,\}$ and $\{\,4, 5, 6, 7, 8\,\}$

c. $\{\,1, 2, 3, 4, 5\,\}$ and $\{\,2, 3\,\}$

d. $\{\,1, 2, 3, 4, 5\,\}$ and $\{\,1, 2, 3, 4, 5\,\}$

3. Match a drawing on the right with the set relationship described on the left:

a. the set of squares; the set of rectangles.　　I. ◯

b. the set of circles; the set of squares.　　II. ◯ ◯

c. the set of right angles; the set of straight angles.　　III. ◯

d. the set of rectangles; the set of trapezoids.　　IV. ◯

4. Draw figures to show the relationships of the given pairs of sets:

a. the set (N) of odd numbers; the set (P) of prime numbers.

b. the set (E) of even numbers; the set (S) of squares of odd numbers.

c. the set (H) of numbers less than 100; the set (D) of squares of whole numbers less than 10.

d. the set (T) of numbers divisible by 2; the set (C) of numbers divisible by 3.

Operations with Sets

In arithmetic, we have the operations of addition, subtraction, multiplication, and division which give us answers to problems. In a similar way, we have operations with sets which give us answers to questions. There are two main operations with sets. These operations are called *union* and *intersection*.

Before we operate with sets, let's recall two things about computing with numbers. Whenever we add or multiply numbers we work with only two numbers at a time. For example, when we add $2+3+4$, we may add $2+3$ to get 5 and then add $5+4$ to get the complete sum, 9. We do the same thing in multiplication. In a similar way, when we work with three sets we operate with only two sets at a time.

Here is another important principle of numbers. The sum of two natural numbers or the product of two natural numbers is always another natural number. For example, $3 + 4 = 7$ and $5 \times 8 = 40$. Mathematicians call this result *closure*. Natural numbers are *closed* with respect to addition or multiplication. In the same way, when we operate with sets, the result is another set.

Unions of Sets

Consider the following two sets:

$$A = \{ \text{Harry, Dale, Bob, Pete} \},$$
$$B = \{ \text{Charles, Mike, Bob, Pete} \}.$$

Suppose we wish to find the set whose members appear in either set A *or* set B or in both A *and* B. This set is:

$$\{ \text{Harry, Dale, Bob, Pete, Charles, Mike} \}.$$

Notice that we don't write the names Bob and Pete twice. As soon as we write them once, we know they are members of the desired set.

This is an example of a *union* operation on two sets. This operation directs us to find the set that includes *all* the members of the sets on which the operation is performed.

Just as we use the symbol $+$ to show the addition of numbers, we use a symbol for union. The symbol for union is $\cup$. It is easy to remember this symbol because it looks like a large capital U. Some people call this symbol "cup" because it is shaped like a cup. We will always call $\cup$ "union." Then, for the example given above,

$$A \cup B = \{ \text{Harry, Dale, Bob, Pete, Charles, Mike} \}.$$

$A \cup B$ is the set of all members of A together with all members of B. The result of a union of two sets is another set, just as the sum of two natural numbers is another natural number. In the same way that $2 + 3$ is a numeral for 5, so $A \cup B$ is a symbol for a set.

For the example above, how many members does set A have? How many members does set B have? How many members does set $A \cup B$ have? Do we add the number of members of each set to get the number of members of the union?

One way to better understand the union of two sets is to make Venn diagrams. We can then shade the regions of these sets to represent the union.

Consider three examples:

a. If $A = \{1, 2, 3\}$ and $B = \{3, 4, 5\}$, then $A \cup B = \{1, 2, 3, 4, 5\}$. This union is represented in Figure 11a.

b. If $C = \{1, 2, 3\}$ and $D = \{4, 5, 6\}$, then $C \cup D = \{1, 2, 3, 4, 5, 6\}$. See Figure 11b.

c. If $X = \{1, 2, 3, 4\}$ and $Y = \{1, 2\}$, then $X \cup Y = \{1, 2, 3, 4\}$. See Figure 11c.

Original sets intersect.   Original sets are disjoint.   One original set is a subset of the other.

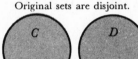

a $A \cup B$ b $C \cup D$ c $X \cup Y$

Figure 11

EXERCISE SET 9
Working with Sets: Unions

1. List the members of the union of each pair of sets:
   a. $A = \{a, b, c, d\}$; $B = \{b, d, f, h\}$
   b. $X = \{2, 4, 6, 8, 10\}$; $Y = \{3, 6, 9, 12\}$
   c. $R = \{\text{Mary, Beth, Liz}\}$; $T = \{\text{Grace, Bonnie, Sue}\}$
   d. $P = \{\triangle, \square, \bigcirc, \star\}$; $Q = \{\triangle, \bigcirc, \square\}$

2. What is the number of elements of each union in Example 1 above?
   a. $n(A \cup B) = ?$ c. $n(R \cup T) = ?$
   b. $n(X \cup Y) = ?$ d. $n(P \cup Q) = ?$

3. Shade the areas that represent these unions:

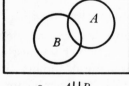

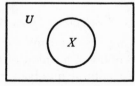

a. $A \cup B$ b. $X \cup U$

158

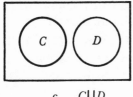

c. $C \cup D$

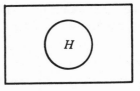

d. $H \cup H$

4. Write the sets that are answers to the following unions:

a. $A \cup A$　　　　b. $A \cup \phi$　　　　c. $A \cup U$

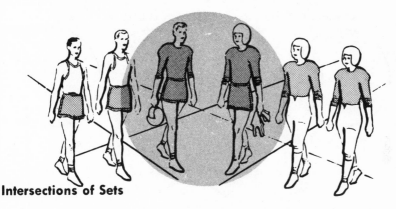

Intersections of Sets

Consider again the sets we used in our discussion of unions:

$A = \{$ Harry, Dale, Bob, Pete $\}$,

$B = \{$ Charles, Mike, Bob, Pete $\}$.

Let us now find the set of boys who belong to *both A and B*. This set is:

$\{$ Bob, Pete $\}$.

We have found the *intersection* of A and B. The intersection operation on two sets directs us to find a set that includes only the common members of *both* sets. The symbol used for intersection is $\cap$. Sometimes this is called "cap." We will always read $\cap$ as "intersection." $A \cap B$ is a symbol for the set of members of set A who are also members of set B. In our example,

$A \cap B = \{$ Bob, Pete $\}$.

Note the difference between "intersecting sets" and an "intersection" of two sets. Intersecting sets are sets that have some members in common. The intersection of two sets is another set consisting of the common members from the two sets. Just as 3×4 is a numeral for 12, so $A \cap B$ is a symbol for a set.

The intersection set $A \cap B$ can be shown by shading regions of Venn diagrams. Consider the following examples:

a. If $A = \{1, 2, 3\}$ and $B = \{3, 4, 5\}$, then $A \cap B = \{3\}$. This intersection is represented by the shaded portion of Figure 12a.

b. If $C = \{1, 2, 3\}$ and $D = \{4, 5, 6\}$, then $C \cap D = \phi$. See Figure 12b.

c. If $X = \{1, 2, 3, 4\}$ and $Y = \{1, 2\}$, then $X \cap Y = \{1, 2\}$. See Figure 12c.

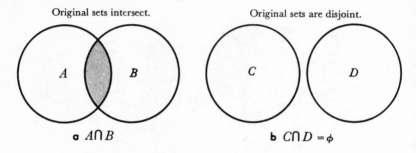

Original sets intersect.   Original sets are disjoint.

a $A \cap B$ b $C \cap D = \phi$

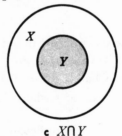

One original set is a subset of the other.

c $X \cap Y$

Figure 12

EXERCISE SET 10
Working with Sets: Intersections

1. List the members of the intersection of each pair of sets:

a. $\{a, b, c, d, e\}$; $\{a, e, i, o, u\}$

b. $\{$ Fords, Chevrolets, Plymouths, Ramblers $\}$; $\{$ Plymouths, Dodges, De Sotos, Chryslers $\}$

c. $\{$ Carl, Bill, Mort, Herb $\}$; $\{$ Charles, Bob, Mark, Hub $\}$

d. the set of prime numbers less than 10; the set of odd numbers less than 10.

2. Copy the drawings and shade the regions that represent the following intersections:

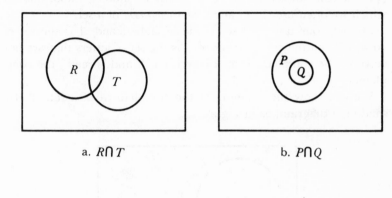

a. $R \cap T$ b. $P \cap Q$

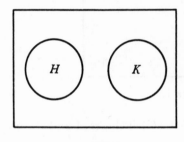   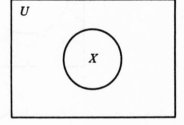

c. $H \cap K$ d. $X \cap U$

3. Write the sets that result from these intersections:

   a. $A \cap A$

   b. $A \cap \phi$

   c. $A \cap U$

4. Data regarding members of sets of students in South High School is as follows:

> $X =$ the set of boys in the ninth-grade mathematics class.
> $Y =$ the set of ninth-grade boys on the football team.
> $n(X) = 12.$
> $n(Y) = 5.$
> $n(X \cap Y) = 3.$

Find $n(X \cup Y)$.

How many football players are in the ninth-grade mathematics class?

5. If we are given $n(X)$, $n(Y)$, and $n(X \cap Y)$, what is the formula for $n(X \cup Y)$?

161

The Complement of a Set

So far we have talked about two special sets — the empty set (ϕ) and the universal set (U). There is another important set which we often use. It is called the *complement* of a set.

Suppose our universal set is the alphabet, and A is the set of vowels. Then the complement of A is the set of letters that are not vowels. The complement of A is written A' and is read "the complement of A" or "not A."

We can show the meaning of the complement of a set, A', by shading a diagram, as in Figure 13.

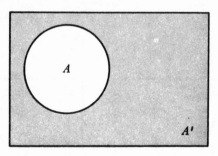

Figure 13

Remember that the universal set (U) includes A and A', or, in other words, $A \cup A' = U$.

Here is another example of a complement. Let our universe be the set of whole numbers and E be the set of even numbers. Then E' is the set of whole numbers that are not even, or, in other words, the set of odd numbers.

EXERCISE SET 11

Working with Sets: Complements

1. Suppose that the universe is the set of natural numbers less than 10, or $U = \{$ 1, 2, 3, 4, 5, 6, 7, 8, 9 $\}$. If $X = \{$ 1, 2, 3, 4, 5 $\}$, what are the members of X'?

2. Suppose that the universe is the set of all automobiles and that A is the set of Cadillacs. Describe A'.

3. If the universe is the set of all triangles and R is the set of right triangles, describe R'.

4. Suppose our universe is Fords, Chevrolets, Plymouths, and Ramblers. If F is the set of all Fords and S the set of all 1960 models, describe in words the sets represented by the following symbols:

a. $F \cap S$ d. $F' \cap S$ f. $F' \cap S'$ h. $(F \cup S)'$

b. $F \cup S$ e. $F \cap S'$ g. $F' \cup S'$ i. $(F \cap S)'$

c. F'

Answering Questions with Sets

The set operations and drawings of set relationships now give us a method for getting answers to many questions, such as, for example, "How many students on the basketball team get an 'A' in mathematics?" In solving problems like this, each member of a set is listed in a Venn diagram according to the conditions known. Suppose we have these sets:

A = the set of boys who get an "A" in mathematics = { Bill, Carl, Jim, Mike, Peter, Gary, Bob, Bud, Barney }.

B = the set of boys on the basketball team = { Tom, Marv, Bill, Carter, John, Charles, Tony, Bob, Barney }.

Since some basketball team members also take mathematics, the circles representing these sets intersect. Our problem is to find $A \cap B$, and to do this we can tabulate the boys in the shaded part of Figure 14.

$A \cap B$

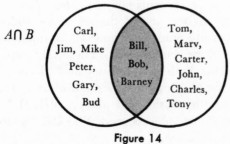

Figure 14

Bill, Bob, and Barney are members of set A and also of set B. These three boys are members of the basketball team and also get "A" in mathematics. The boys who are members of set A or set B can easily be seen in the figure.

More difficult questions, involving three sets, can also be answered by this method of analysis. Suppose we have three clubs, A, B, and C, and ask questions such as "How many girls belong to only one club?" and "How many girls belong to three clubs?"

Let $A = \{$ Carol, Ann, Sue, Barb $\}$,
$B = \{$ Jane, Carol, Judy, Barb $\}$,
$C = \{$ Mary, Sue, Carol, Judy $\}$.

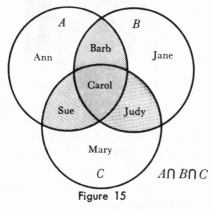

Figure 15

Figure 15 shows that Carol belongs to all three sets; Ann, Jane, and Mary belong to only one set; while Judy, Sue, and Barb belong to two sets.

Now consider a still more difficult problem. In a certain school there are 21 students in a mathematics course, 17 in physics, and 10 in advanced history. Of these, there are 12 students taking both mathematics and physics, 6 taking both mathematics and history, and 5 taking both physics and history; but these figures include 2 students who take all three subjects. If the three classes are combined for a field trip, accommodations will have to be secured for how many different students?

Let X be the set of all students taking mathematics, Y the set of all students taking physics, and Z the set of all students taking history. Since there are students taking all combinations of the three subjects, the three sets intersect. Then $X \cap Y$ represents the set of students taking both mathematics and physics, $X \cap Z$ the set of students taking both mathematics and history, $Y \cap Z$ the set of students taking both physics and history, $X \cap Y \cap Z$ the set of students taking all three subjects, and $X \cup Y \cup Z$ the total set of students under consideration.

Look now at the Venn diagram in Figure 16.

Region G represents the students who take mathematics, physics, and history.

$$n(G) = n(X \cap Y \cap Z)$$
$$= 2$$

Region F represents those who take physics and history, but no mathematics.

$$n(F) = n(Y \cap Z) - n(G)$$
$$= 5 - 2$$
$$= 3$$

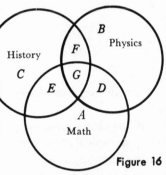

Region E represents those who take mathematics and history, but no physics.

$$n(E) = n(X \cap Z) - n(G)$$
$$= 6 - 2$$
$$= 4$$

Region D represents those who take physics and mathematics, but no history.

$$n(D) = n(X \cap Y) - n(G)$$
$$= 12 - 2$$
$$= 10$$

Figure 16

Region C represents those who take history, but no mathematics or physics.

$$n(C) = n(Z) - [n(F) + n(G) + n(E)]$$
$$= 10 - (3 + 2 + 4)$$
$$= 1$$

Region B represents those who take physics, but no mathematics or history.

$$n(B) = n(Y) - [n(F) + n(G) + n(D)]$$
$$= 17 - (3 + 2 + 10)$$
$$= 2$$

Region A represents those who take mathematics, but no physics or history.

$$n(A) = n(X) - [n(D) + n(E) + n(G)]$$
$$= 21 - (10 + 4 + 2)$$
$$= 5$$

The total number of different students, $n(X \cup Y \cup Z) = n(A) + n(B) + n(C) + n(D) + n(E) + n(F) + n(G) = 5 + 2 + 1 + 10 + 4 + 3 + 2 = 27$, so we need accommodations for 27 students.

Figure 17 shows the number of students in each region. Now we also know how many students have each combination of courses.

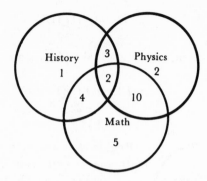

Figure 17

This is the way sets can be used to find the answers to many questions about groups of persons or events.

On several occasions we have mentioned problems involving chance or probability. Let us now consider a typical problem. Suppose we wish to find what our chances are of getting a peppermint gumdrop from a vending machine that has just been filled with 50 cherry gumdrops, 25 licorice gumdrops, and 75 peppermint gumdrops.

Set ideas can be used to solve this problem. Let C be the set of cherry gumdrops, L the set of licorice gumdrops, and P the set of peppermint gumdrops. Then $n(C) = 50$, $n(L) = 25$, and $n(P) = 75$. The Venn diagram in Figure 18 shows that all three sets are disjoint.

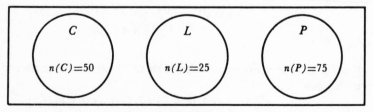

Figure 18

The chance or probability of an event happening (such as getting a peppermint gumdrop) is defined as the number of *favorable* ways the event could happen, divided by the *total* number of ways the event could happen.

The number of ways you could receive a peppermint gumdrop from the machine is equal to the number of peppermint gumdrops in the machine and is $n(P)$ or 75 ways. The total number of ways you could receive a gumdrop is $n(C \cup L \cup P) = 50 + 25 + 75 = 150$. Therefore the chance of getting a peppermint gumdrop from a filled machine is $\dfrac{75}{150} = \frac{1}{2}$, or, in other words, your chances are 1 out of 2.

EXERCISE SET 12
Solving Problems with Sets

1. Draw Venn diagrams and tabulate the members in regions to show how the following pairs of sets are related:
   a. $\{ a, b, c, d, e \}$ and $\{ a, e, i, o, u \}$
   b. $\{ 1, 2, 3, 4, 5, 6 \}$ and $\{ 4, 5, 6, 7, 8 \}$
   c. $\{$ Peggy, Jean, Sue $\}$ and $\{$ Margy, Kay, Sue $\}$
   d. $\{$ John, Bob, Charles, Don $\}$ and $\{$ John, Bob, Charles $\}$

2. Draw Venn diagrams and tabulate members in regions to show how the following groups of sets are related:
   a. $\{ a, b, c, d \}$; $\{ b, d, e, f \}$; $\{ a, b, e, g \}$
   b. the set of even natural numbers less than 10, the set of prime numbers less than 10, the set of natural numbers less than 10 divisible by 2 or 3.
   c. $S =$ the set of boy scouts $= \{$ Tom, Dick, Harvey, Mert $\}$
      $R =$ the set of boys in seventh-grade mathematics $= \{$ Jim, Harry, Mert, Carl, Rob $\}$
      $T =$ the set of boys delivering papers $= \{$ Dick, Harry, Rob, Bill $\}$

3. A survey of the seventh grade gives this data for TV viewing:
$$60\% \text{ see program } A$$
$$50\% \text{ see program } B$$
$$50\% \text{ see program } C$$
$$30\% \text{ see program } A \text{ and } B$$
$$20\% \text{ see program } B \text{ and } C$$
$$30\% \text{ see program } A \text{ and } C$$
$$10\% \text{ see program } A, B, \text{ and } C$$

   a. What percent view A and B but not C?
   b. What percent view exactly two programs?
   c. What percent do not view any program?

4. Consider the following data about a universe and several sets of numbers:

$$U = \{\ 1, 2, 3, 4, 5, 6, 7, 8, 9, 10\ \}$$
$$X = \{\ 2, 4, 6, 8\ \}$$
$$Y = \{\ 1, 3, 5, 7\ \}$$
$$Z \cap Y = \{\ 1, 3\ \}$$
$$Z \cup Y = \{\ 1, 2, 3, 5, 7, 9\ \}$$

Use Venn diagrams to find:

a. the numbers in set Z

b. $n(X) - n(Z \cap X)$

c. $n(Z \cup X) - n(Z \cup Y)'$

5. On a certain day, three sports writers picked the following teams to win games in the American League:

{ New York, Chicago, Detroit, Boston }

{ New York, Cleveland, Washington, Detroit }

{ Boston, New York, Baltimore, Washington }

Kansas City was not picked to win by any of the writers.

Use Venn diagrams to determine the American League teams that played each other on that day.

6. A drawing for prizes is conducted by putting 15 blue cards (representing the cheapest prizes), 5 red cards (representing more expensive prizes), and one white card (for the grand prize) into a box.

a. If you draw first, what is your chance of getting a red card?

b. If you draw first, what is your chance of getting either the white card or a blue card?

Laws of Operation with Sets

Did you ever learn the laws of operation to follow when computing with numbers? Or did you just learn the number combinations without worrying about the rules? In arithmetic and algebra, there are three laws for computation:

1. **Commutative law,**

    or law of order: **Arithmetic**   **Algebra**

The order in which $2 + 3 = 3 + 2$ $a + b = b + a$
we add two numbers $5 \times 4 = 4 \times 5$ $ab = ba$
or multiply two num-
bers does not affect
the sum or product.

2. **Associative law,**

    or law of grouping:

The way in which we $(2 + 3) + 4 = 2 + (3 + 4)$ $(a + b) + c = a + (b + c)$
group quantities to $(4 \times 5) \times 6 = 4 \times (5 \times 6)$ $(ab)c = a(bc)$
be added or multi-
plied does not affect
the sum or product.

3. **Distributive law:**

The multiplier of a $2(3 + 4) = (2 \times 3) + (2 \times 4)$ $a(b + c) = ab + ac$
sum multiplies each
addend of the sum.

These laws help us check our computations and simplify many formulas in algebra.

Since there are basic laws of operation in arithmetic and algebra, we might suspect that there are similar principles for set operations, too.

Suppose we have these three sets:

       Tabulation   **Venn Diagram**

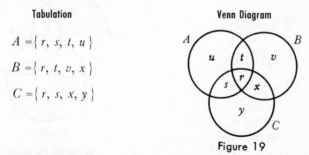

$A = \{\, r,\, s,\, t,\, u \,\}$

$B = \{\, r,\, t,\, v,\, x \,\}$

$C = \{\, r,\, s,\, x,\, y \,\}$

Figure 19

Let us see if we can apply an associative law to the union operation on these sets. In other words, does $(A \cup B) \cup C = A \cup (B \cup C)$?

1. Find $A \cup B$.

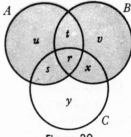

Figure 20

By tabulation
$A \cup B = \{ r, s, t, u \} \cup \{ r, t, v, x \}$
$A \cup B = \{ r, s, t, u, v, x \}$

By shading the diagram
$A \cup B = \{ r, s, t, u, v, x \}$

2. Find $(A \cup B) \cup C$.

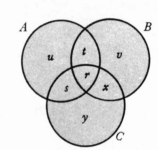

Figure 21

By tabulation
$(A \cup B) \cup C = \{ r, s, t, u, v, x \} \cup \{ r, s, x, y \}$
$(A \cup B) \cup C = \{ r, s, t, u, v, x, y \}$

By shading the diagram
$(A \cup B) \cup C = \{ r, s, t, u, v, x, y \}$

3. Find $A \cup (B \cup C)$.

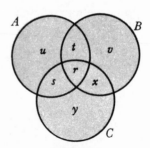

Figure 22

By tabulation
$B \cup C = \{ r, s, t, v, x, y \}$
$A \cup (B \cup C) = \{ r, s, t, u, v, x, y \}$

By shading the diagram
$A \cup (B \cup C) = \{ r, s, t, u, v, x, y \}$

From the tabulations and Venn diagrams, we see that $(A \cup B) \cup C = A \cup (B \cup C)$.

Let us now use the same sets to consider an associative law applied to intersections. In other words, does $(A\cap B)\cap C = A\cap(B\cap C)$?

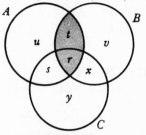

1. Find $A\cap B$.

Figure 23

By tabulation
$A\cap B = \{r, s, t, u\}\cap\{r, t, v, x\}$
$A\cap B = \{r, t\}$

By shading the diagram
$A\cap B = \{r, t\}$

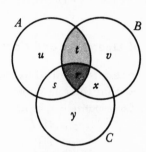

2. Find $(A\cap B)\cap C$.

Figure 24

By tabulation
$(A\cap B)\cap C = \{r, t\}\cap\{r, s, x, y\}$
$(A\cap B)\cap C = \{r\}$

By shading the diagram
$(A\cap B)\cap C = \{r\}$

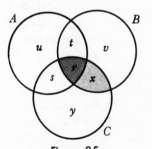

3. Find $A\cap(B\cap C)$.

Figure 25

By tabulation
$B\cap C = \{r, x\}$
$A\cap(B\cap C) = \{r\}$

By shading the diagram
$A\cap(B\cap C) = \{r\}$

From the tabulations and Venn diagrams, we see that $(A\cap B)\cap C = A\cap(B\cap C)$.

From these results it seems that the associative law or the law of grouping holds for the unions and intersections of sets:

1. *Unions:* $(A \cup B) \cup C = A \cup (B \cup C)$
2. *Intersections:* $(A \cap B) \cap C = A \cap (B \cap C)$

EXERCISE SET 13

Discovering Laws of Operations for Sets

Using the sets A, B, and C of the preceding section, draw Venn diagrams and tabulate regions to show that:

1. $A \cup B = B \cup A$
2. $A \cap B = B \cap A$
3. $A \cap (B \cup C) = (A \cap B) \cup (A \cap C)$
4. $A \cup (B \cap C) = (A \cup B) \cap (A \cup C)$

Laws of Operation and Computing Machines

The results obtained in Exercise Set 13 and the section preceding it show that these laws of operation hold for sets:

1. Commutative law, or law of order:
$$A \cup B = B \cup A$$
$$A \cap B = B \cap A$$

2. Associative law, or law of grouping:
$$(A \cup B) \cup C = A \cup (B \cup C)$$
$$(A \cap B) \cap C = A \cap (B \cap C)$$

3. Distributive law:
$$A \cup (B \cap C) = (A \cup B) \cap (A \cup C)$$
$$A \cap (B \cup C) = (A \cap B) \cup (A \cap C)$$

We know that there are many applications of these laws in arithmetic and algebra. Where do we need to know these laws of operation in sets? One dramatic application of these laws is their use in solving problems with electronic computers. Suppose we wanted to find out what factors are related to success in playing basketball. Sets of data about height and weight, eating habits, amount of rest, running speed, and coordination would be gathered. Then relationships between these sets of data would be represented by unions and intersections. These relationships would then show how the electric circuits in the computer should be arranged so that the computer could give us the answers we're interested in finding.

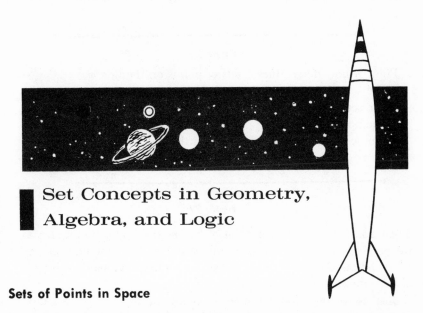

Set Concepts in Geometry, Algebra, and Logic

Sets of Points in Space

Will we be able to send a space ship to the moon? Would you like to travel in space? To do so, you will need to know something about geometry. And the study of the geometry of space involves many ideas about sets.

To navigate in space you will need to know about points, lines, planes, and three-dimensional objects. Let's begin with a point. What is a point? We may think of a point as being a location in space. We represent it with a dot like a period (.). But a point has no shape or size or weight. A point is really an idea; nobody has ever seen a true point. In fact, mathematicians do not even give a definition for a point, and thus we say it is an *undefined term*.

In space geometry, we talk about sets of points. In fact, we say that space is an infinite set of points. Similarly, we think of a line as being an infinite set of points, a set of certain points. Any point on this line is a member of this set of points. In geometry, we talk about many types of lines, such as straight lines, curved lines, and broken lines, but when we use the term *line*, here, we will be talking about a straight line. A part of a straight line like *A*_____*B* is called a *line segment*. The set of points on a line segment is also an infinite set. Every line, short or long, straight or curved, is an infinite set of points.

Let's take a point on a line such as point *A* on line *m* in Figure 26.

A

m

Figure 26

Point *A* now determines 3 sets of points on the line *m:*
  the set of points on *m* to the left of point *A;*
  the set of points on *m* to the right of point *A;*
  the point *A* itself.

Suppose another line *n* crosses *m* at the point *A*.

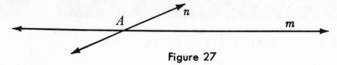

Figure 27

Then point *A* is the intersection of line *m* and line *n*. Point *A* is a member of the set of points on line *m* and is also a member of the set of points on line *n*. This agrees with our previous description of the intersection ($\cap$) of sets.

Just as we talk about the set of points on a line, we can talk about the set of lines on a point (lines passing through a point) such as point *A* in Figure 28.

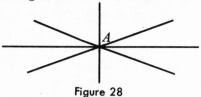

Figure 28

Is the set of lines on a point an infinite set?

Figure 29 indicates how we can show a one-to-one correspondence between infinite sets — the set of lines on point *A*, the set of points on line *m*, and the set of points on line *n*. Point *A* and any point on line *m* can be connected by a line. This line will pass through a point on line *n*. Hence, for every point on *m* there is a corresponding point on *n* and a corresponding line through *A*.

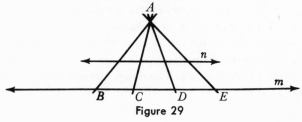

Figure 29

A plane surface like a table top is also a set of points. It is the set of points representing a flat surface. A plane has two dimensions — length and width — but it has no thickness and so is invisible, just as points and lines are. The set of points on a plane, the set of lines on a plane, the set of planes passing through a line (planes on a line) are all infinite sets.

Mathematicians say a line on a plane divides the plane into two *half planes*. In Figure 30 a line *m* on a plane divides the plane into three sets of points: the set of points, *A*, in the half plane above the line; the set of points, *B*, in the half plane below the line; and the set of points on the line. Are these sets all infinite sets?

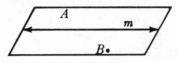

Figure 30

A closed curve also divides the plane into sets of points, as shown in Figure 31. A simple closed curve like a circle or a triangle divides a plane into three sets of points: the inside, the outside, and the boundary. To go from a point on the inside to a point on the outside, the path must cross the boundary.

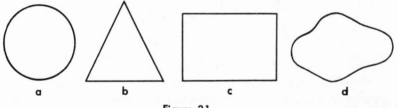

Figure 31

The curves in Figure 32 are also closed curves, but they divide the plane into more than 3 sets of points. These curves are not simple closed curves.

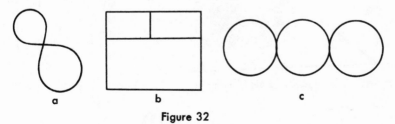

Figure 32

When a line crosses a closed curve on a plane, the points of crossing are the members of the intersection set of the two sets.

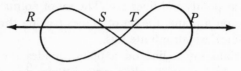

Figure 33

In Figure 33, the intersection set (I) of the closed curve and line m has the points R, S, T, P as members. $I = \{R, S, T, P\}$.

In three-dimensional space, a plane divides space into three sets of points: the set of points in the half space above the plane, the set of points in the half space below the plane, and the set of points on the plane. A closed surface like a balloon or a cube divides space into three sets of points: the inside, the outside, and the boundary. To go from a point on the inside to a point on the outside, the path must cross the boundary. The objects in Figure 34 are examples of simple closed surfaces.

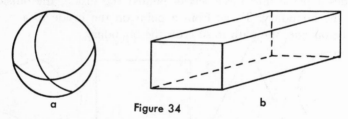

a   **Figure 34**   b

When two planes intersect, the intersection set is a line. When a plane and a closed surface intersect (see Figure 35), the intersection set may be a point, a line, or a set of lines.

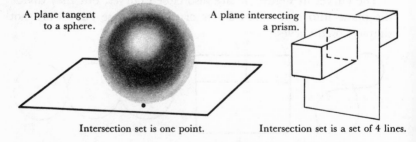

A plane tangent to a sphere.   A plane intersecting a prism.

Intersection set is one point.   Intersection set is a set of 4 lines.

Figure 35

EXERCISE SET 14
Sets and Geometric Figures

1. Make a drawing, wherever possible, to show the relationship between a straight line and a circle if the intersection set between the two is:

    a. the empty set   c. a set of two members

    b. a set of one member   d. a set of three members.

2. Into how many sets of points do these figures divide the plane?

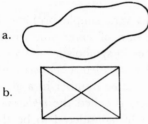

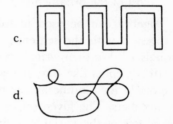

a.   c.

b.   d.

3. Make a sketch of a circle and a quadrilateral for which the intersection set is:

    a. a set of 2 points   c. a set of 6 points

    b. a set of 4 points   d. a set of 8 points.

4. Refer to the figure on the right for the following exercises:

    a. What is the intersection set of the rectangle and line *m*?

    b. Copy the figure and shade the intersection set of the set of points inside the rectangle and the set of points on the half plane which contains *C*.

    c. Copy the figure and shade the intersection set of the set of points outside the rectangle and the set of points on the half plane which does not contain *C*.

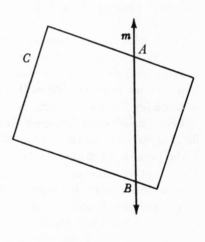

5. Consider the intersection set of a sphere and a plane. Which of the following intersection sets are possible?

    a. the empty set   c. a set with two members

    b. a set with one member   d. a set of points on a simple closed curve

Sets and Sentences

We have seen how set concepts can be applied to the field of geometry; now we shall examine the use of sets in the study of algebra. Let's start with an easy problem:

The \$7.00 selling price of a unique table ornament from France is made up of the actual cost of the ornament plus \$3.00 in taxes and import duties. What is the actual cost of the ornament?

Algebra can be used to solve this very simple problem if we express it with the sentence, "The sum of some number and 3 equals 7." We then write this sentence in symbols as: $x + 3 = 7$.

In sentences like this, the symbol x is called a *variable* and is a *placeholder* for a numeral, the name of some number in a given set of numbers. The given set of numbers is called the *replacement set* or *domain* of the variable and is usually considered to be the set of all real numbers. Usually we want to find replacements for x that will make the sentence true. The numbers from the replacement set which make a sentence true are members of the *solution set* for the sentence. The solution set for the sentence $x + 3 = 7$ has only one member, $x = 4$. If $x = 4$, then we have the true statement, $4 + 3 = 7$.

Some solution sets of sentences have two members; for example, the solution set for $x^2 = 9$ is $+3$ and -3.

What replacement for x makes the sentence $x + 5 = x$ true? We know of no number that will make this a true statement. The solution set is the empty set.

Let's consider the sentence $3x + x = 2x + 2x$. What replacement for x makes this sentence true? The solution set for this sentence is the infinite set of all numbers. It is called an *identity* since it is always a true sentence.

In a similar way, we often need solution sets for "inequations" or inequalities. A sentence such as "a number is greater than 3" is written $x > 3$. Does replacing x with 7 make this sentence true? Is $7 > 3$? The set of numbers that makes this a true sentence has an infinite number of members. If $x = 5$ or 7 or 19 or $325\frac{1}{2}$, the sentence $x > 3$ is true. All these numbers, 5, 7, 19, $325\frac{1}{2}$, are members of the solution set for $x > 3$.

If, however, we say the universe for the sentence $x > 3$ is the set of natural numbers less than 10, then the solution set for $x > 3$

is $\{4, 5, 6, 7, 8, 9\}$. In this way, equations and inequalities are called *set selectors:* they select from a universe of numbers just those numbers that make the sentence true when used as replacements for x.

We have said that in algebra we use a letter such as x as a placeholder for numerals. In a similar way, we may use x as a placeholder for members of a set. The sentence "the set of all numbers x such that x is a prime number" is abbreviated this way: $\{x|x$ is a prime number$\}$. The vertical bar | is read "such that." The set of all numbers x such that x is less than 5 is written $\{x|x<5\}$. The symbol $\{x| \quad \}$ is called the *set builder.*

EXERCISE SET 15
Finding Solution Sets

Find the solution sets of integers for these set selectors:
1. $\{x|3x+7=22\}$
2. $\{x|x^2=25\}$
3. $\{x|x>5\}$ where $U=\{1, 2, 3, 4, 5, 6, 7, 8\}$
4. $\{x|x+2=x\}$
5. $\{x|3x+5=2x+5+x\}$

Pictures of Solution Sets

Consider again the sentence $x>3$. Since the solution set for this inequality is an infinite set, we cannot list all the members. But the solution set for $x>3$ can be represented pictorially by a straight line. Since a line has an infinite number of points on it, the points will have a one-to-one correspondence with the members of the solution set. The elements of the set $\{x|x>3\}$ are all numbers greater than 3. It does not include 3. The shaded line of Figure 36 represents the solution set of $x>3$ and is called the graph of that solution set. The rounded arrowhead indicates that 3 is not included in the graph.

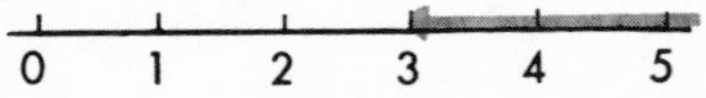

Figure 36

If the sentence had been written $x \geq 3$, then 3 would have been included in the set. The symbol $\geq$ is read "equal to or greater than." The sentence $x \geq 3$ is read, "Find all x such that x is equal to or greater than 3."

We sometimes encounter inequality problems with two parts; for example, "Find the number or numbers greater than 5 or less than 2." This problem can be expressed by the "compound sentence" $x > 5$ or $x < 2$. (Note that the symbol $<$ means "less than.") Since the solution set for $x > 5$ *or* the solution set for $x < 2$ satisfies the conditions of the compound sentence, the union of these two sets will be the solution set for the compound sentence.

Let $A = \{ x | x < 2 \}$, $B = \{ x | x > 5 \}$, and $C = \{ x | x < 2 \ or \ x > 5 \}$. Then $C = A \cup B$.

The graph of $x < 2$ is shown in Figure 37,

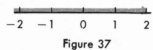

Figure 37

and the graph of $x > 5$ is given in Figure 38.

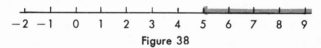

Figure 38

Figure 39 shows the graph of $(x < 2) \cup (x > 5)$.

Figure 39

If we restrict the universe so that $U = \{-1, 0, 1, 2, 3, 4, 5, 6, 7, 8, 9\}$, then $C = \{ x | x < 2 \ or \ x > 5 \} = A \cup B = \{ -1, 0, 1, 6, 7, 8, 9 \}$.

If the set selector had been $\{ x | x > 2 \ and \ x < 5 \}$ the solution set would have been the intersection of the sets of numbers that satisfy the conditions $x > 2$ and $x < 5$.

The graphs are shown in Figure 40.

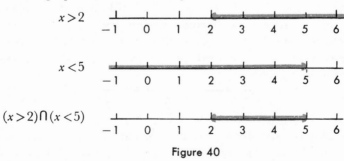

Figure 40

1. Graph the solution sets for these sentences:
   a. $x + 5 = 9$ c. $x > \frac{1}{2}$
   b. $x < -3$ d. $x + 2x = 3x$

2. Graph these compound sentences:
   a. $\{ x | x = 3 \text{ or } x > 3 \}$ c. $\{ (x < 2) \cup (x > 6) \}$
   b. $\{ x | x \geq 3 \text{ and } x < 7 \}$ d. $\{ (x > 5) \cap (x < 9) \}$

Sets of Ordered Pairs

See if you can solve this problem: "If you double the amount of money that Harry has, Harry and Joe together have 7 dollars."

If we let the variable x hold a place for the number of dollars Harry has, and y hold a place for the number of dollars Joe has, then the problem can be expressed by the sentence $2x + y = 7$. We have used two placeholders in this sentence. What are some values of x and y that will make this sentence true? One pair is $x = 2$, $y = 3$. Are there other pairs of values for x and y that will make this sentence true?

When a sentence has two variables (usually designated x and y for the sake of convenience), the members of the solution set are pairs of numbers. The sentence $2x + y = 7$ can be regarded as a selector of a set of number pairs that make the sentence true. If the replacement set for x and y is the set of all real numbers, then the solution set of $2x + y = 7$, denoted by $\{ (x, y) | 2x + y = 7 \}$, is an infinite set of number pairs. The symbol $\{ (x, y) | 2x + y = 7 \}$ is read: "The set of ordered pairs (x, y) such that $2x + y = 7$."

It is desirable to have a simple notation for number pairs. We write a pair like $x = 2$, $y = 3$ as $(2, 3)$, with the x value first and the y value second. Since the order is significant, we call $(2,3)$ an *ordered pair*. The pair $(3,2)$ is not the same as $(2,3)$ and cannot be used as a replacement for it. Note that the pair $(3,2)$ does not make $2x + y = 7$ a true sentence, for $2 \times 3 + 2$ does not equal 7.

Working with Ordered Pairs

1. Find 3 members of the set of ordered pairs that make these sentences true. Consider the replacement set as the set of all real numbers.
 a. $x + y = 5$
 b. $2x + y = 12$
 c. $3x - y = 7$

2. If the replacement set for x and y is $U = \{-2, 2, 4, 6\}$, find an ordered pair for each of the following sentences that will make them true:
 a. $x - y = 6$
 b. $3x - 2y = 10$

Sets of Ordered Pairs and Operations

Sometimes sets of ordered pairs are used to show definite relationships between two sets of numbers. For example, the set of ordered pairs $\{(1, 6), (3, 8), (0, 5) \ldots\}$ shows that the second member of each pair is obtained by adding 5 to the first member. Similarly, $\{(2, 10), (3, 15), (4, 20), (0, 0) \ldots\}$ shows that the second member of each pair is obtained from the first member by multiplying it by 5.

Mathematicians call any set of ordered pairs a *relation*. A relation such that for every value of x there is only *one* value of y is called a *function*. The above two sets of ordered pairs are functions. The set of number pairs $\{(1, 1), (1, 2), (1, 3) \ldots\}$ is not a function, for there is more than one value of y for a certain value of x.

EXERCISE SET 18
Relations, Functions, and Ordered Pairs

1. Determine the operation, if any, used to obtain the second member of each number pair from the first:
 a. $\{(9, 4), (7, 2), (21, 16) \ldots\}$
 b. $\{(9, 3), (6, 2), (2, \frac{2}{3}) \ldots\}$
 c. $\{(0, 0), (3, 3), (7, 7) \ldots\}$
 d. $\{(5, 6), (9, 10), (84, 85) \ldots\}$
 e. $\{(1, 3), (2, 5), (1, -1), (2, -4)\}$

2. Which of the relations of Exercise 1 are functions?

Ordered Pairs and the Cartesian Plane

In several previous problems, we used a number to locate a point on a line. If we work with a pair of lines that intersect at right angles and form (determine) a plane surface, we need a pair of numbers to locate a point.

Let us consider a universe of three numbers, $U = \{1, 2, 3\}$, and form the set of all possible ordered pairs: $\{(1, 1), (1, 2), (1, 3), (2, 1), (2, 2), (2, 3), (3, 1), (3, 2), (3, 3)\}$. This set is denoted by $U \times U$, read "U cross U."

The lattice of points in Figure 41 represents the set of number pairs obtained from $U = \{1, 2, 3\}$, and is called the graph of $U \times U$. The point indicated by the shaded dot represents the number pair $x = 1$, $y = 3$, and is said to have the *coordinates* (1, 3). Note that there is a one-to-one relationship between every point in the graph and the set $U \times U$.

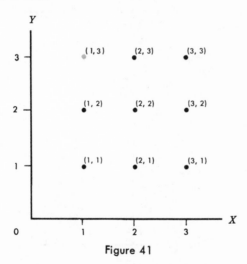

Figure 41

A set of number pairs $U \times U$ is called the Cartesian set of U, a name honoring René Descartes, a famous French mathematician. Descartes showed the relationship of algebra and geometry by associating number pairs with points in a plane. If U refers to the set of all real numbers, then the graph of $U \times U$ is an infinite geometric plane called the *Cartesian plane*.

We can use the Cartesian plane to graph sentences with 2 placeholders. The ordered pairs which make the sentence true will be the coordinates of the points that make up the graph.

Consider again the sentence $2x + y = 7$. If the replacement set for x and y is $U = \{1, 2, 3, 4, 5, 6, 7\}$, then the solution set must be found in the set $U \times U$, and $\{(x, y) | 2x + y = 7\} = \{(1, 5), (2, 3), (3, 1)\}$. The graph of the solution set is shown in Figure 42a.

If U is the set of all real numbers, the graph of the solution set of $2x + y = 7$ becomes a continuous set of points (a straight line) as shown in Figure 42b.

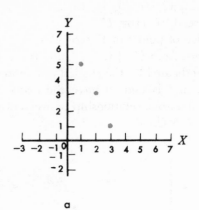

a

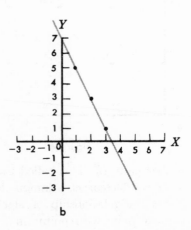

b

Figure 42

We can also use the Cartesian plane to graph inequality sentences having two variables, such as $x > y$. If the solution set must be found in $U \times U$, where $U = \{1, 2, 3\}$, then, by inspection,

$\{ (x, y)|x>y \} =\{ (2, 1),\ (3, 1),(3, 2) \}$. Figure 43*a* shows the graph of the solution set to be three points.

If U is the set of all real numbers, the graph of the solution set of $x>y$ can be found by drawing the straight line graph of the solution set of the sentence $x =y$. In this case, every point *below* the line has a y coordinate with a value *less* than the value of its x coordinate. The y coordinate of every point *above* the line is greater than the x coordinate of that point. Thus the shaded part of Figure 43*b* represents the infinite solution set of $x>y$.

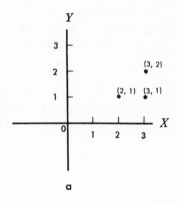

a

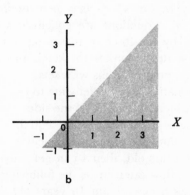

b

Figure 43

1. Draw the graph of $U \times U$ if:
   a. $U = \{ 1, 3, 5 \}$
   b. $U = \{ -2, 0, 2 \}$

2. Draw the graph of the solution set of $x + y = 3$ if the replacement set for x and y is:
   a. $U = \{ 1, 2, 3 \}$
   b. the set of all real numbers.

3. Draw the graph of the solution set of $y > x$ if the replacement set for x and y is:
   a. $U = \{ 1, 2, 3 \}$
   b. the set of all real numbers.

4. Draw the graph of the solution set for $y < 2x + 3$ if the replacement set for x and y is the set of all real numbers.

5. A certain company uses two substances, x and y, to make a cleaning compound. When the two substances are mixed, the x substance must exceed the y substance by more than 3 gallons, but the combined volume of x and y must be less than 7 gallons. Use graphing methods to find the greatest number of gallons of substance y that can be used at one time.

Sets and Logic

Are all high school students impolite? Are some beautiful girls intelligent? Are boy scouts ever juvenile delinquents? Conclusions like these can be tested by the use of sets.

One of the best uses for sets is their use in making logical deductions. Logical deductions are conclusions obtained from certain assumptions by the force of reasoning. One way of using sets to make logical deductions is to sketch the relationship as we did in the section on operations with sets.

In mathematics, most statements to be proved are written in an "if-then" fashion, like this: "If the sides of a triangle are all equal, then the angles of the triangle are equal."

We can state a non-mathematical situation the same way: "If I am 16 years old, then I can get a driver's license." Another way to make this statement is the following: "All 16-year-olds can get a driver's license. I am 16 years old. Therefore, I can get a driver's license." Let's represent this situation with sets. Let $A =$

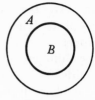

Figure 44

the set of all persons eligible for a driver's license, and B = the set of all 16-year-olds. The relationship between sets A and B is shown in Figure 44. In the drawing, the larger circle represents set A, and the smaller circle represents set B.

The drawing tells us that every member of set B is a member of set A and is therefore eligible for a driver's license.

Suppose we make this statement: "If all girls are beautiful creatures and all wives are girls, then all wives are beautiful creatures." This statement can be written in three sentences:

(*a*) All girls are beautiful creatures.

(*b*) All wives are girls.

(*c*) Therefore, all wives are beautiful creatures.

We would represent these statements in sets as follows:

A = the set of all beautiful creatures.

B = the set of all girls.

C = the set of all wives.

Figure 45 shows how we would represent these relationships in a drawing.

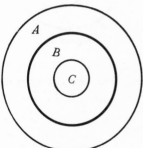

Figure 45

In set notation, the analysis is stated this way:

$A \cap B = B$	All girls (B) are included in set A and hence are beautiful creatures.
$B \cap C = C$	All wives (C) are included in set B and hence are girls.
$A \cap C = C$	All wives (C) are included in set A and hence are beautiful creatures.

Both the drawing and the set analysis show that the conclusion, "All wives are beautiful creatures," is a logical conclusion.

Statements like this, called *syllogisms*, have been considered ever since the days of the Greek philosophers. They are made up of three parts or statements, in this order:

(*a*) the *major premise* ("All girls are beautiful creatures.")

(*b*) the *minor premise* ("All wives are girls.")

(*c*) the *conclusion* ("All wives are beautiful creatures.")

Every syllogism has three *terms*, in this case, "beautiful creatures," "girls," and "wives." Each term occurs twice in the syllogism.

The statements made in a syllogism can be classified as either *universal* statements or *particular* statements.

A universal statement includes *all* members of the sets mentioned in the statement. This statement can be an affirmative statement such as "All girls are beautiful creatures." The sets involved have the relationship shown in Figure 46. The one set is a subset of the other (or they could be identical).

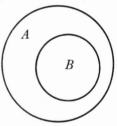

Figure 46

The universal statement could be a negative, such as "no girls are beautiful creatures." The sets involved in a universal negative statement are disjoint, as shown in Figure 47.

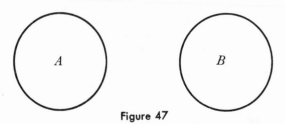

Figure 47

A particular statement involves only *some* members of a set. "Some girls are beautiful" is an example of an affirmative particular statement. Figure 48 shows that this set relationship can be pictured in either of two ways. The sets could intersect, indicating that some girls are not beautiful creatures and some beautiful creatures are not girls, or the set of beautiful creatures could be just a subset of the set of girls.

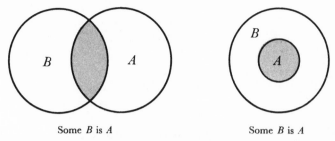

Some *B* is *A*   Some *B* is *A*

Figure 48

The negative particular statement might say, "Some girls are not beautiful." Notice that the set diagrams shown in Figure 49 are the same as the diagrams of Figure 48.

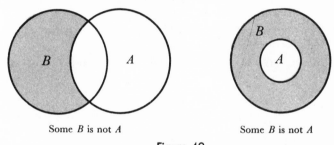

Some *B* is not *A*   Some *B* is not *A*

Figure 49

Since a particular statement can lead to two possible set relationships, it is more difficult to make a logical conclusion based on a particular statement than on a universal statement.

Many syllogisms have both universal and particular statements; for example: "If all hi-fi record players have several speakers and some radios are also hi-fi record players, then some radios have several speakers." This syllogism can be analyzed as follows:

(*a*) All hi-fi record players have several speakers.

(*b*) Some radios are also hi-fi record players.

(*c*) Some radios have several speakers.

Let *A* = the set of all sound-reproducing instruments with several speakers.

    B = the set of hi-fi record players.

    C = the set of radios.

The diagram in Figure 50 shows this relationship.

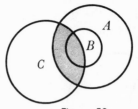

Figure 50

The diagram shows that the conclusion follows from the premises. It is impossible for C to intersect B without also intersecting A.

The set analysis of the problem looks like this (the symbol $\neq$ means "not equal"):

$A \cap B = B$	All B is A
$B \cap C \neq \phi$	Some C is B
$A \cap C \neq \phi$	Some C is A

Here is a sample syllogism with negatives.

(*a*) All mathematicians have long hair.

(*b*) Some politicians have short hair (do not have long hair).

(*c*) Some politicians are not mathematicians.

The shaded area of Figure 51 shows that the conclusion is logical.

L = the set of all long-haired people.

M = the set of mathematicians.

P = the set of politicians.

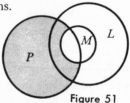

Figure 51

In all of our syllogism examples, we have reached conclusions from certain major and minor premises. The conclusions were logical, but were not necessarily true, for some of our premises were unacceptable assumptions. We must realize that conclusions obtained by the force of reasoning are no better than the premises or assumptions from which they are deduced. We can only be sure that logical conclusions are true if we start with acceptable premises. Nevertheless, logic is an important useful tool in modern society, and the fact that set ideas can be easily applied to logic makes the study of sets all the more practical.

1. Represent these statements about sets by drawings. Label each drawing to indicate the sets involved.

    a. All even numbers are whole numbers.

$$E = \{ \text{ even numbers } \}$$
$$W = \{ \text{ whole numbers } \}$$

    b. No odd numbers are even numbers.

$$O = \{ \text{ odd numbers } \}$$
$$E = \{ \text{ even numbers } \}$$

    c. Some boys are intelligent.

$$B = \{ \text{ boys } \}$$
$$I = \{ \text{ intelligent people } \}$$

2. Copy these sketches and then label each sketch so that it will show the proper relationship between the sets involved in the following syllogisms:

    a. (*a*) All boys are interested in mathematics.

       (*b*) All "A" mathematics students are boys.

       (*c*) All "A" mathematics students are interested in mathematics.

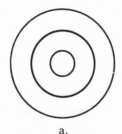

a.

    b. (*a*) No students who own a car get "A" grades.

       (*b*) Some girls get "A" grades.

       (*c*) Some girls do not own a car.

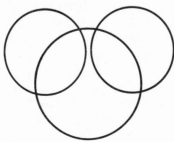

b.

c. (a) All students at St. Mark's school are boys.
  (b) Some courteous students are in St. Mark's school.
  (c) Some boys are courteous.

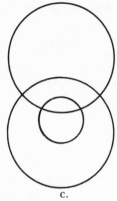

c.

d. (a) All girls are honest.
  (b) No politicians are honest.
  (c) No girls are politicians.

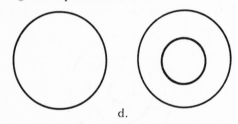

d.

3. Make sketches to represent these syllogisms. Decide whether or not each conclusion is logical.

    a. (a) All metals are elements.
      (b) All iron is a metal.
      (c) Iron is an element.
    b. (a) All freshmen are 15 years old.
      (b) No 15-year-olds are intelligent.
      (c) No freshmen are intelligent.

4. Test each of the proposed conclusions to see if they are logical by using diagrams.

Major premise: All timid creatures are bunnies.
Minor premises: Some timid creatures are dumb.
                Some freshmen are timid creatures.
Conclusions. Some bunnies are dumb.
                Some freshmen are bunnies.
                Some freshmen are dumb bunnies.

A Backward Glance and a Look to the Future

We have seen how simple ideas about sets or collections of objects have become a very important part of the world of mathematics, how sets can be used to simplify, clarify, and solve a variety of problems, and how set concepts are applied to geometry, algebra, and logic. However, we have been able to examine only a few aspects of the mathematical study of sets. Many mathematicians believe that the theory of sets is so simple and so basic that it can be used to unify the study of mathematics.

As you continue your work in mathematics, look for ways to apply set concepts to new problems. Your textbooks may not use set notation, but the set *ideas* are more important than the notation. Mastering these ideas can enable you to take a big step toward becoming adept in mathematics, the study of structures with patterns that are like " . . . lace, the leaves of trees, and the play of light and shadow. . . ."

EXERCISE SET 21
Review Test on Sets

A. Which of the following statements are true?
1. Every set has at least one member.
2. The empty set is a subset of all whole numbers.
3. If two sets have the same number of members, the sets are equal.
4. There is a one-to-one correspondence between the atoms in the earth and the natural numbers.
5. There are as many members in the set of natural numbers as in the set of numbers divisible by 17.
6. If A is the set of vowels and B is the set of the letters of our alphabet, then $A \subset B$.
7. The set of all boys and the set of all girls are disjoint.
8. If $P \subset Q$, then $Q' \subset P'$.
9. If $n(A) = n(B)$, then A is equivalent to B.
10. There is a one-to-one correspondence between the points on line xy and the line RT. x_____y R_____T

B. Copy and shade these drawings to show the solution sets of the indicated relations.

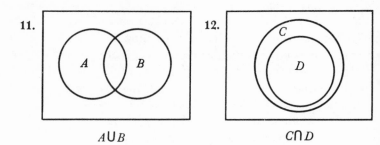

11.

$A \cup B$

12.

$C \cap D$

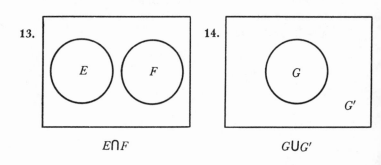

13.

$E \cap F$

14.

$G \cup G'$

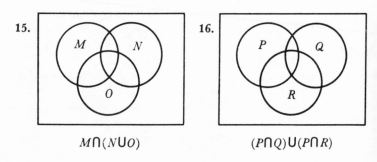

15.

$M \cap (N \cup O)$

16.

$(P \cap Q) \cup (P \cap R)$

C. Sketch diagrams to represent these relationships between sets:

  17. the set (A) of numbers divisible by 3 and
  the set (B) of numbers divisible by 7.

  18. the set (C) of odd numbers and
  the set (D) of squares of odd numbers.

  19. $X = \{ 1, 2, 3, 4, 5 \}, Y = \{ 2, 4, 6, 8 \}, Z = \{ 3, 4 \}$

  20. Find the members of the set $X \cap (Y \cup Z)$, using sets X, Y, and Z of item 19.

D. Given $U = \{ 1, 2, 3, 4, 5, 6, 7, 8, 9 \}$
  $\qquad A = \{ 1, 2, 3, 4, 5 \}$
  $\qquad B = \{ 2, 4, 6, 8 \}$

  21. $A \cap B =$

  22. $B' =$

  23. $A' \cap B =$

  24. $A \cup B =$

  25. $A \cap \phi =$

Extending Your Knowledge

Now that you have an understanding of sets, you may want to go further in your study of this important mathematical topic. Here are some books that will help you extend your knowledge.

ADLER, IRVING, *The New Mathematics*. The John Day Co., 1958

AIKEN, DAYMOND J., and BESEMAN, CHARLES A., *Modern Mathematics: Topics and Problems*. McGraw-Hill Book Co., 1959

ALLENDOERFER, C. B., and OAKLEY, C. O., *Principles of Mathematics*. McGraw-Hill Book Co., 1955

BANKS, J. HOUSTON, *Elements of Mathematics*. Allyn and Bacon, Inc., 1956

KEMENY, JOHN G., SNELL, J. LAURIE, and THOMPSON, GERALD L., *Introduction to Finite Mathematics*. Prentice-Hall, Inc., 1957

WOODWARD, EDITH J., and McLENNAN, RODERICK C., *Elementary Concepts of Sets*. Henry Holt and Co., 1959

PART V

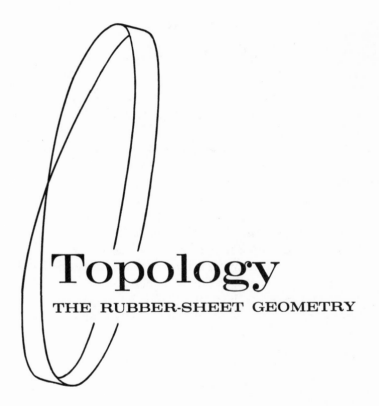

Topology
THE RUBBER-SHEET GEOMETRY

The Strange World of Topology

What Is Topology?

Have you ever heard of a sheet of paper with only one side? Why do mathematicians say that a doughnut and a flowerpot are more alike than a doughnut and a pretzel? When is a triangle the same as a circle? Is it possible to change a left shoe to a right shoe by taking a trip around space? These are the kinds of questions that topology answers. This doesn't sound very much like mathematics, does it? But it is, and it's one of the newest and most exciting fields of mathematics. Since it talks about things that are familiar to you, like the inside of a glove or the difference between right and left shoes, it will not be too strange for you. And topology is so full of impossibilities, tricks, and puzzles that it will be fun to learn more about it.

Topology is the branch of mathematics that decides what is possible. It tells us whether it is possible to turn an inner tube inside out. You may think this is an easy problem. Topologists say it is possible, but no one has ever been able to do it with a real inner tube.

In topology, we never ask, "How long?", "How far?", or "How big?" Instead, we ask, "Where?", "Between what?", "Inside or outside?" A traveler on a strange road wouldn't ask, "How far is Centerville?" if he didn't know the direction. The answer, "Three miles from here," would not help him very much if there were several roads. He is more likely to ask, "How do I get to Centerville?" Then the answer, "Follow this road until you come to a fork, then turn to your left," will tell him how to get to Centerville. This answer doesn't sound mathematical because it says nothing about distances and does not describe whether the path is straight or curved. This is the kind of answer that topology gives to questions.

opology and Geometry

Topology is something like geometry because it deals with lines, points, and figures. But the figures are different from those of geometry because they are permitted to change in size and shape. So, someone has called it "rubber-sheet geometry." Topology is more interested in position than in size or shape. It deals with the properties of position that are not affected by changes in size and shape. For example, suppose you draw a square on a rubber sheet, with a dot inside the square. No matter how you stretch the rubber sheet, the dot will always be on the inside of the square. So topology is the study of geometric properties that stay the same in spite of stretching or bending.

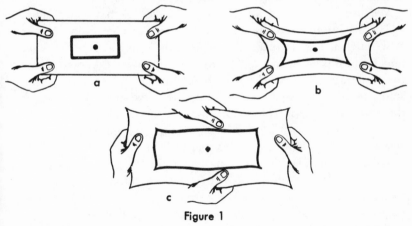

Figure 1

Distance has no meaning in topology. Two points an inch apart may easily be made two inches apart by stretching. Likewise, angle size is meaningless because you can stretch a rubber sheet so that an angle of 15° becomes an angle of 35°. Even straight lines have no meaning in topology because the straight line AB

$$A \text{———————} B$$

Figure 2

may become a curved line like the following by stretching the sheet:

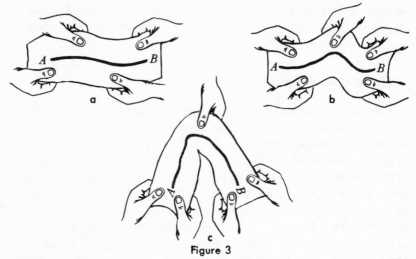

Figure 3

The straight line not only becomes a curved line, but it also changes in length.

We usually think of an object like a key as hard and rigid. It keeps its shape and fits the lock for years and years, no matter how much it has been moved about. When an airplane takes off and flies away it seems to become smaller. But we know it stays the same size no matter where it is. Euclidean geometry is the study of objects which always stay the same size. Topology is the study of things which do change in size and shape when moved. It starts with the idea that there are no rigid bodies; everything can change in size, shape, and position.

We can think of a line as being like a piece of string. If a point is on a line, like a knot on a string. it must remain on that line even though the line is twisted, stretched, or curved in many ways. We also say that a line is *continuous*. There are no holes in the line. Whenever a line crosses another line, it passes through a point on that line. This means, for example, that if you draw a line *CD* through line *XY*, as in the figure below, line *CD* passes through a point on line *XY*.

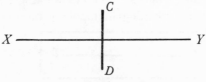

Figure 4

So many of the properties of lines and figures change in this rubber-sheet geometry that you may think nothing remains the same. This is not true. Look at the line *AB* in Figure 2 again. No matter how we stretch or bend the sheet, the path from *A* to *B* remains a path from *A* to *B* which does not cross itself. The line or path may become crooked, or longer, even more than in Figure 3, but it remains a line or path from *A* to *B*. In topology, a path or line like *AB* is called *Arc AB*.

How Geometric Figures Change in Topology

What we have said about simple lines like *AB* also applies to lines that form geometric figures such as circles or triangles.

Let's see what happens to a circle on a rubber sheet. By stretching the sheet, the circle may change as pictured below.

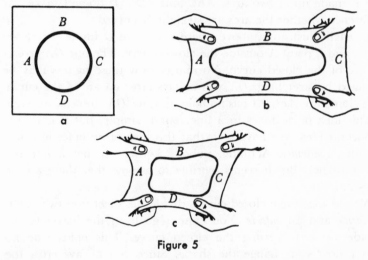

Figure 5

We can see that the circle changes a great deal in shape and size. But no matter how we stretch the sheet, the figure remains a path, *ABCDA*. We can also see that no matter where we start on this path, we will return to the starting point without crossing the path. If we start at *C*, we will pass through *B*, *A*, *D* and return to *C*. In topology, all these figures have the same name. Each is called a *simple closed curve* or a *closed circuit*. Each is made up of the two arcs *ABC* and *ADC* which have only the points *A* and *C* in common.

Look at the geometric figures below:

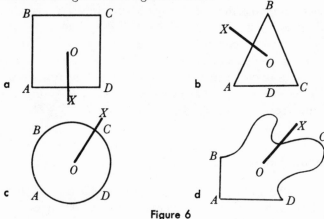

Figure 6

Topology says these figures are all simple closed curves. Each one is made up of two arcs, ABC and ADC. It doesn't make any difference whether the arcs are straight or curved.

In each illustration above, there is a point O inside the closed curve and a point X outside the closed curve. The line OX crosses an arc of the closed curve. No matter how these figures may be changed by stretching, OX will always cross an arc of the curve. The closed arc $ABCDA$ has no holes in it for OX to sneak through.

This idea of no holes in a line sounds simple, but it is a very important idea. We have seen that this idea of no holes in a line is called *continuity*. Actually, nobody knows whether a line has a hole or not. But it seems sensible to assume that there are no holes in a line.

We say that these closed curves divide the sheet into two parts, an *inside* and an *outside*. You cannot go from the inside to the outside without crossing the closed curve. This holds true no matter how you change the shape. Since you always cross the line in going from the outside to the inside no matter how you distort the figure, we call this crossing an *invariant* situation. Any situation in topology that stays the same under distortion is called an *invariant*. When we distort a figure — for example, a straight line stretched into a curved line or a square into a circle — we have made a change called a *topological change* or *topological transformation*. These transformations change the size or shape of the figure but do not form a new topological figure. If we cut, tear, or fold a line or a surface, we change the line or surface so that it has new features. So a topological transformation is made *without* cutting, tearing, folding, or punching holes.

In the circle *ABCD*, another property which does not change, or is invariant, is the order of the points *A*, *B*, *C*, *D*. What was invariant about the line *AB*? No matter how it was stretched, it remained a path from *A* to *B* without crossing itself. We have seen that in topology a circle may change into an ellipse or a square and a straight line may become a curved line. But when we join points *A* and *B* of the line *AB*, as in Figure 7, we have a new figure, a closed curve.

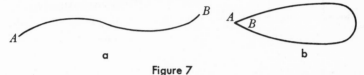

Figure 7

In a similar way, when we cut the arc of a circle, as in Figure 8, we change the closed curve to a line.

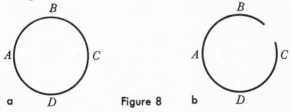

Figure 8

These changes are *not* transformations. New topological figures are formed.

In geometry, we study the properties of size, shape, area, and angle size. We say figures are congruent when we can place one figure upon another of the same shape and size with all parts matching. Topological transformations give figures that are said to be *equivalent*. In topology, the circle and square are equivalent no matter how they differ in size. Both of them have one inside and one outside. To go from the inside to the outside, we must cross one line. If we shade the inside of the figure below we can easily see how it divides the surface of the picture into two regions.

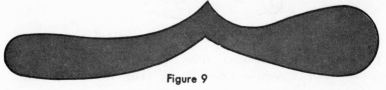

Figure 9

The Persian Caliph and His Daughter's Boy Friends

The idea of an inside and an outside helps solve interesting problems, like the old story about a Persian Caliph who used a topology problem to select a husband for his beautiful daughter. She had so many boy friends that he decided to pick the one who was the best problem solver. The first problem given to the

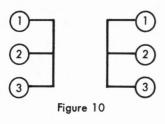

Figure 10

boy friends is illustrated in Figure 10. The problem was to connect like numbers by lines that did not cross each other or any other lines in the figure. Any boy who solved this problem successfully could then talk to the Caliph's daughter.

This was an easy problem and made all the boys excited. See if you can solve this problem. Would you have been permitted to talk to the Caliph's daughter?

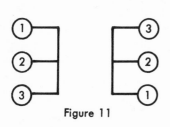

Figure 11

But the boy who could marry the Caliph's daughter had to solve a second problem. The second problem was again to connect like numbers with lines that did not cross each other or any other lines. But note how the drawing has changed. Can you solve this problem?

Someone has said the Caliph's daughter died an old maid. What do you think?

A solution for the first problem looks like this.

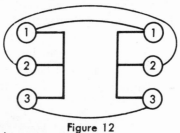

Figure 12

That was easy!

For the second problem, let's draw lines from 1 to 1, and 2 to 2. Now we have a simple closed curve. The inside is shaded.

One 3 is inside the closed curve, and the other 3 is outside. And we know that you can't get from the inside to the outside of a simple closed curve without crossing a line; so topology says it is impossible to draw the lines without crossing.

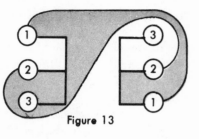

Figure 13

In the Caliph's problem, we worked with a simple closed curve having one inside and one outside. Topology is also concerned with other curves that are not simple closed curves. Closed figures like the one below have more than one inside. Do you see five regions (four inside areas) in Figure 14?

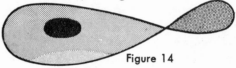

Figure 14

Now look at Figure 15. In each drawing, into how many regions is the sheet divided? How many inside regions does each one have?

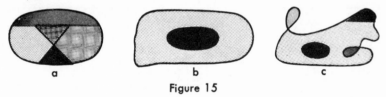

a   b   c

Figure 15

EXERCISE SET 1
Topological Curves and Regions

1. Which of the following figures are topological lines?

a   b   c   d

2. Which of the following figures are simple closed curves?

a   b   c   d

3. How many inside regions does each of these figures have?

a   b   c   d

Fun with the Moebius Strip

All the closed surfaces we have talked about have been on the surface of a sheet. All the sheets we know of, like a sheet of paper, have two surfaces — a front and a back. Have you ever seen a piece of paper with only one surface? There really is such a sheet. It is called a *Moebius strip* and has been used by many magicians to entertain people. It has been a plaything for mathematicians ever since it was discovered by August Ferdinand Moebius, a German mathematician, in 1858. A fly can walk from any point on this strip to any other point without crossing an edge. Unlike a sheet of paper or a table top, it does not have a top or a bottom, or a front or a back.

You can make a Moebius strip with any strip of paper. Any size or type of paper will do, but gummed tape an inch or two wide and one or two feet long is easy to handle. We use the strip to make a ring or band. But before we glue the ends together, we give one end a half-twist. If you use gummed tape, twist one end so that you stick the gummed side of one end to the gummed side of the other end. Attach the band as illustrated in Figure 16.

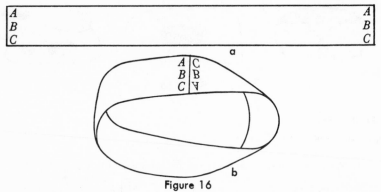

Figure 16

If you draw a line on the surface of your Moebius strip, you will find that you will go all around the entire surface without crossing an edge. Paint or color one surface without going over an edge. Is there another surface that remains to be colored?

For another unusual result, cut the band lengthwise along a line in the center of the strip. What unexpected result did you obtain? If you make another band, and cut it lengthwise one third of the way in from an edge such as at point B in Figure 16, you will get still a different result.

If you put the Caliph's problem that we discussed on page 8 on a Moebius strip, you should be able to win the Caliph's daughter. Try it. In doing this, be sure to draw your lines all the way around the strip, in the same way that you painted the strip.

The Moebius strip enables us to take a new look at right- and left-handed objects like shoes or gloves. If you compare the two gloves of a pair of gloves, you will find that they are equal in all measurements you can make. But you know the gloves are very different. The left-handed glove won't fit your right hand.

How can you change a right-handed glove to a left-handed one? In two dimensions it seems possible on a Moebius strip. If you could slide a picture of the glove along the surface of a Moebius strip, the glove would be upside down and backward when it got back to the starting point.

EXERCISE SET 2
Moebius Strip Facts

You can have fun showing your friends the odd results you get by cutting Moebius strips in different ways. Copy and complete the table below to see what happens when you change the number of twists and the way in which you cut the strip.

Number of Half-Twists	Number of Sides and Edges	Kind of Cut	Results of Cut (Number of sides and edges, length and width, number of loops, twists, and knots)
0		center	
1		center	
1		one-third	
2		center	
2		one-third	
3		center	
3		one-third	

Here's a story about the legendary Paul Bunyan that shows how the Moebius strip has some "practical" applications.

Paul Bunyan versus the Conveyor Belt

By William Hazlett Upson

One of Paul Bunyan's most brilliant successes came about not because of brilliant thinking, but because of Paul's caution and carefulness. This was the famous affair of the conveyor belt.

Paul and his mechanic, Ford Fordsen, had started to work a uranium mine in Colorado. The ore was brought out on an endless belt which ran half a mile going into the mine and another half mile coming out — giving it a total length of one mile. It was four feet wide. It ran on a series of rollers, and was driven by a pulley mounted on the transmission of Paul's big blue truck "Babe." The manufacturers of the belt had made it all in one piece, without any splice or lacing, and they had put a half-twist in the return part so that the wear would be the same on both sides.

After several months' operation, the mine gallery had become twice as long, but the amount of material coming out was less. Paul decided he needed a belt twice as long and half as wide. He told Ford Fordsen to take his chain saw and cut the belt in two lengthwise.

"That will give us two belts," said Ford Fordsen. "We'll have to cut them in two crosswise and splice them together. That means I'll have to go to town and buy the materials for two splices."

"No," said Paul. "This belt has a half-twist — which makes it what is known in geometry as a Moebius strip."

"What difference does that make?" asked Ford Fordsen.

"A Moebius strip," said Paul Bunyan, "has only one side, and one edge, and if we cut it in two lengthwise, it will still be in one piece. We'll have one belt twice as long and half as wide."

"How can you cut something in two and have it still in one piece?" asked Ford Fordsen.

Paul was modest. He was never opinionated. "Let's try this thing out," he said.

They went into Paul's office. Paul took a strip of gummed paper about two inches wide and a yard long. He laid it on his desk with the gummed side up. He lifted the two ends and brought them together in front of him with the gummed sides down. Then he turned one of the ends over, licked it, slid it under the other end, and stuck the two

*Reprinted from the *Ford Times*, by permission of the Ford Motor Company.

gummed sides together. He had made himself an endless paper belt with a half-twist in it just like the big belt on the conveyor.

"This," said Paul, "is a Moebius strip. It will perform just the way I said — I hope."

Paul took a pair of scissors, dug the point in the center of the paper and cut the paper strip in two lengthwise. And when he had finished — sure enough — he had one strip twice as long, half as wide, and with a double twist in it.

Ford Fordsen was convinced. He went out and started cutting the big belt in two. And, at this point, a man called Loud Mouth Johnson arrived to see how Paul's enterprise was coming along, and to offer any destructive criticism that might occur to him. Loud Mouth Johnson, being Public Blow-Hard Number One, found plenty to find fault with.

"If you cut that belt in two lengthwise, you will end up with two belts, each the same length as the original belt, but only half as wide."

"No," said Ford Fordsen, "this is a very special belt known as a Moebius strip. If I cut it in two lengthwise, I will end up with one belt twice as long and half as wide."

"Want to bet?" said Loud Mouth Johnson.

"Sure," said Ford Fordsen.

They bet a thousand dollars. And, of course, Ford Fordsen won. Loud Mouth Johnson was so astounded that he slunk off and stayed away for six months. When he finally came back he found Paul Bunyan just starting to cut the belt in two lengthwise for the second time.

"What's the idea?" asked Loud Mouth Johnson.

Paul Bunyan said, "The tunnel has progressed much farther and the material coming out is not as bulky as it was. So I am lengthening the belt again and making it narrower."

"Where is Ford Fordsen?"

Paul Bunyan said, "I have sent him to town to get some materials to splice the belt. When I get through cutting it in two lengthwise I will have two belts of the same length but only half the width of this one. So I will have to do some splicing."

Loud Mouth Johnson could hardly believe his ears. Here was a chance to get his thousand dollars back and show up Paul Bunyan as a boob besides. "Listen," said Loud Mouth Johnson, "when you get through you will have only one belt twice as long and half as wide."

"Want to bet?"

"Sure."

So they bet a thousand dollars and, of course, Loud Mouth Johnson lost again. It wasn't so much that Paul Bunyan was brilliant. It was just that he was methodical. He had tried it out with that strip of gummed paper, and he knew that the second time you slice a Moebius strip you get two pieces — linked together like an old fashioned watch chain.

Topology Solves
Some Interesting Problems

A Bridge Problem and Topology

In the eighteenth century, in the sleepy German university town of Koenigsberg (now the Russian city of Kaliningrad), Sunday strollers were fond of ambling along the banks of the Preger River, which meandered through the town and was crossed by seven bridges. These bridges ran from each bank of the river to two islands in the river, with one bridge joining the islands, as shown in this drawing.

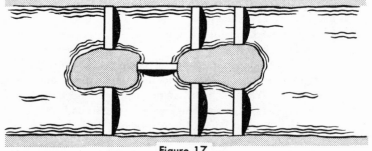

Figure 17

One day, a native asked his neighbor this question: "How can you take a Sunday stroll so that you cross each of our seven bridges and cross each bridge exactly once?" The problem intrigued the neighbor and soon caught the interest of many other people of Koenigsberg as well. They pondered the question seriously, but no one could come up with an answer.

Somehow, the problem came to the attention of a Swiss mathematician by the name of Leonhard Euler, who was serving at the court of the Russian empress Catherine the Great in St. Petersburg. Euler focused his mathematical skill on the problem and eventually came up with a solution: the bridges could not be crossed in the manner posed by the problem. Perhaps the man who first raised the question was disappointed, but he may have been less unhappy had he known that, in the process of working out the problem, Euler had founded the branch of mathematics which we are now examining, topology.

Traveling Networks

In solving the Koenigsberg bridge problem, Euler didn't find it necessary to go to Koenigsberg. He remained in St. Petersburg and did what mathematicians usually do: he drew a diagram of the problem. With this diagram, the land and shore became points, and the bridges became lines connecting these points, as in Figure 18 below. Now you can work on the problem with a pencil, as Euler did. See if you can draw the figure in Figure 18*b* by starting at some point and returning to that point without retracing any line or lifting your pencil off the paper. With this drawing, Euler invented networks and discovered relationships that have been very valuable in topology.

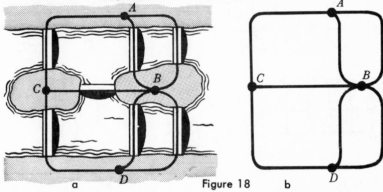

a   **Figure 18**   b

The diagram of the Koenigsberg bridge problem is called a *network*. The points where the lines cross are called *vertices*, and the lines representing bridges are called *arcs*. A network is *traveled* or *traced* by passing through all the arcs exactly once. You may pass through the vertices any number of times. In the network of the Koenigsberg bridges, the vertices are *A*, *B*, *C*, and *D*. The number of arcs to vertex *A* is 3, so the vertex at *A* is called an *odd vertex*. In the same way, *B* is an odd vertex, since 5 arcs go to this vertex. Euler discovered that there must be a certain number of odd vertices in any network if you are to travel it in one journey without retracing any arcs. Euler also discovered other important laws for traveling networks. Maybe you can discover the same ones by trying the exercises below. Copy the geometric figures. Then study the vertices and travel the networks to see if you can discover the relationships between vertices of closed networks, as Euler did.

EXERCISE SET 3
Traveling Networks Experiment

For each network, tabulate the number of even vertices and the number of odd vertices, and then see if the network can be traveled. Copy and complete the table that follows the illustrations

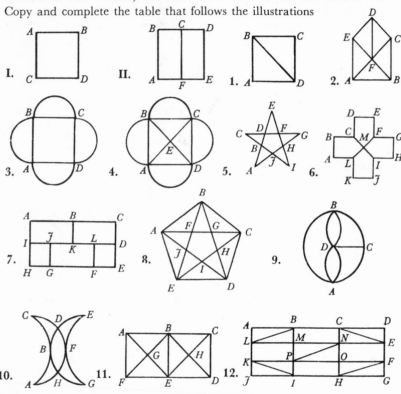

Network	Even Vertices	Odd Vertices	Can it be traveled?
I.	4	0	Yes
II.	4	2	Yes
1.			
2.			
3.			
4.			
5.			

Network	Even Vertices	Odd Vertices	Can it be traveled?
6.			
7.			
8.			
9.			
10.			
11.			
12.			

Euler's Discoveries About Networks

Many of the problems of topology are related to networks. Study the networks in Exercise Set 3 to find the answers to these questions about the relations of vertices and arcs:

1. Where does an arc always begin or end?
2. If two arcs meet, what is true about their vertices?
3. Are all parts of the network connected by arcs?
4. Can a network be traveled in one journey if it has only 2 odd vertices?
5. Can a network be traveled in one journey if it has more than 2 odd vertices?
6. Can a network be traveled in one journey if it has only odd vertices?
7. Can a network be traveled in one journey if it has only 2 even vertices?
8. Can a network be traveled in one journey if it has more than 2 even vertices?
9. Can a network be traveled in one journey if it has all even vertices?

By finding the answers to questions like these, Euler made four general discoveries about networks. First, he showed that the number of odd vertices in a network is always even if it is to be traveled in one journey. Try to travel a network with an odd number of odd vertices in one journey. If you succeed, you will be making mathematical history!

Next, Euler found that a network of all even vertices could be traveled in one journey. In other words, we could start from any vertex, cover the entire network, and return to the same vertex without traveling any arcs twice.

Euler's third discovery was that if a network contained 2, and only 2, odd vertices it could be traveled in one journey, but it would be impossible to return to the starting point. It would be necessary in this case to start at one odd vertex and end at the other.

Euler's last discovery was that if a network contained 4, 6, 8, or any higher even number of odd vertices it would be impossible to travel the network in one journey. In these cases the number of journeys required would be equal to half the number of odd vertices.

Let's see how this works with the Koenigsberg bridges. This network consists of 4 odd vertices, which means that it is impossible to travel the network in one trip. In fact, Euler's fourth discovery shows that it would take exactly two trips.

While Euler was working with the Koenigsberg bridge problem he realized that he was working with a new kind of geometry. He could see that the pattern did not depend on the size of the figure or the shape of the figure. Out of these ideas grew the branch of mathematics now called topology. The networks we have studied have *not* been concerned with length, area, angles, or shape. Instead, the important thing has been places, and how the places are connected by arcs. In geometry, we study the properties of figures that remain the same when you move a figure without changing its shape. For example, a circle has a certain radius, diameter, circumference, and area which does not change if it is moved from one place to another. When we move figures in geometry the motion is rigid; that is, we do not allow any change in shape. In topology, we can move figures and change their shape by twisting or stretching, forgetting about length or distance, angles, and arcs. In topology, we study the properties of the figure that remain the same under this distortion.

Now let us see how networks apply to the curves we have discussed earlier. The simplest network is a single arc such as AB: A _____B. A and B are said to be *vertices* of this

single arc. Another simple network is A ⟨⟨⟨⟩⟩⟩ B, the closed

curve with 2 vertices and 2 arcs. However, we may locate other points on A B, so that it may have many vertices, such as:

A ⟨⟨⟨⟩⟩⟩ B . Examples of complex networks are the lines
$\quad C \quad D$
$\quad E \quad F \quad G$

on a basketball court, or the cities and roads on a road map.

EXERCISE SET 4
Network Problems and Puzzles

1. The three houses below, A, B, and C, must each connect to the water main, W, the gas main, G, and the electric line, E. Is it possible to make these connections so that no lines cross? Draw a network to help you decide whether or not it is possible. Maybe shading the drawing, as we did in the Caliph's problem, will help you find the answer to this puzzle.

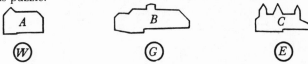

2. Is it possible to take an entire trip through a house whose floor plan is shown in the figure below and pass through each door once and only once? Try drawing a network to correspond to this figure. Let the rooms be vertices and the doors be arcs.

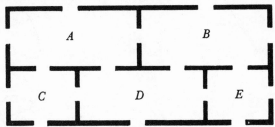

Networks, Regions, and an Important Formula

One of the questions asked about networks is the number of pieces, or regions, into which a network divides a plane surface. For example, in Figure 19a below there is only 1 region. You can go from any point in the plane to any other point in the plane without crossing the network.

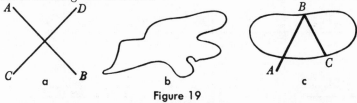

Figure 19

How many regions are there for a closed curve like the one in Figure 19b? Can you count 4 regions for the network of Figure 19c?

What relationships exist between regions, arcs, and vertices? If you study the networks in Exercise Set 5 you may be able to discover the relationship between the number of vertices (V), the number of arcs (A), and the number of regions (R) of a network.

EXERCISE SET 5

Networks and Regions Experiment

Copy and complete the table that follows the illustrations for each of the networks and then see whether you can state a formula relating the three variables, V, A, and R.

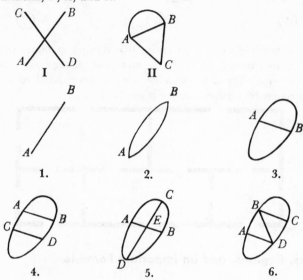

Network	V = the Number of Vertices	A = the Number of Arcs	R = the Number of Regions
I	5	4	1
II	3	4	3
1.			
2.			
3.			
4.			
5.			
6.			

Euler's Formula for Networks

If you are a mathematical wizard you would discover that
$$V - A + R = 2.$$

This is Euler's network formula and expresses one of the most important properties of networks. See if it works for the networks in Figure 20.

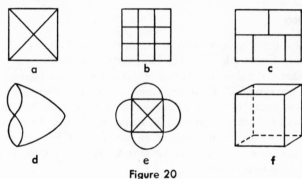

Figure 20

The Four-Color Map Problem

One of the most famous unsolved problems in mathematics which is related to networks and regions is the four-color map problem. Suppose we want to draw a map so that countries with a common border are colored differently. How many different colors are needed to make this kind of map?

The drawings below illustrate some possible maps. Copy these maps and color the countries so that countries with common borders have different colors.

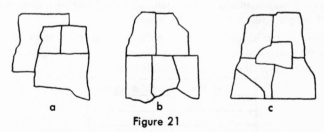

Figure 21

So far, it has been possible to color all maps that can be thought of using only four different colors. However, it has never been proved mathematically that four colors are enough for all maps. It has been proved that five colors are sufficient to color any

217

map, and that to color some maps it is necessary to have at least four colors. But mathematicians are still searching for a proof to show that four colors are sufficient as well as necessary. So if you want to become famous, find a map that needs five colors or prove that four colors are sufficient.

EXERCISE SET 6
Maps and Colors

1. Perform these exercises with a map of the United States.
   a. Draw a network of six states in your region.
   b. What is the smallest number of states that you can pass through in traveling from Portland, Maine, to San Francisco, California?
   c. How many different colors are needed to color the map of part *b* so that all states in that region are colored differently?
   d. What state or states border on the greatest number of other states, when "border on" means a common border which is more than a point?

2. Copy this map. Color it with four colors so that no states with common borders have the same color.

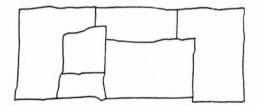

3. Draw a map with ten regions that can be colored with three colors, no two bordering regions having the same color.

4. Divide a circle into regions with lines so that no three lines go through the same point. What is the largest number of regions formed for each number of lines? Copy and fill in this table.

No. of Lines	No. of Regions
1	
2	
3	
4	
5	

How do the differences in number of regions compare for each cut? How many regions would you predict for six lines?

A Topological Look at Our Three-Dimensional World

Classifying Topological Figures

In our space age, of course, we should consider how topology applies to three-dimensional objects, such as spheres, cubes, and doughnut-shaped objects that occupy space.

In topology, a sphere is something like a circle. It divides space into one inside region and one outside region. A cube or a pyramid does the same. To get from a point inside these figures to the outside, the path or line crosses the surface of the object in one point. Again we assume there is no hole in the surface of the object, just as we say there is no hole in a line. This means a surface is continuous. Any closed surface which divides space into two regions, an inside and an outside, is a *simple closed surface*. So a sphere, a cube, and a pyramid are simple closed surfaces. Any simple closed surface like a cube or a pyramid can be changed to a sphere by distortion.

What is the difference between a simple closed curve and a ring or between a sphere and a doughnut? Topology says it is how the lines or surfaces are connected.

And topology tells us how to change or transform one figure or object into a different figure or object. For example, consider a ring on a flat surface, as in Figure 22.

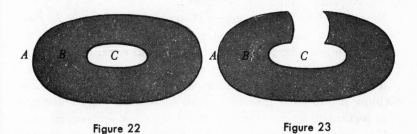

Figure 22   Figure 23

This ring figure is not a simple closed curve. It divides the sheet into three regions, *A*, *B*, and *C*. But if we cut the ring once, as in Figure 23, it becomes a simple closed curve. *A* and *C* are now both on the outside.

Topology classifies objects or figures by finding out how many cuts are needed to simplify the figure or surface. A ring like the one above is classified as a *singly* connected surface because it takes one cut to transform it to a simple closed curve. Be sure to notice that the words *singly* and *simple* mean different things in classifying figures.

In three dimensions, topology classifies an object according to the number of cuts necessary to change it to a simple closed surface like the sphere. For example, the doughnut is somewhat like the ring we described above. If we cut across a doughnut once, as shown in Figure 24, it becomes a simple closed surface that can be distorted into a sphere. Note that a single cross-cut of a simple closed surface would produce two pieces. Thus the result produced by a single cut is the primary topological distinction between a doughnut and a sphere. We say that the doughnut and the sphere differ in "connectivity."

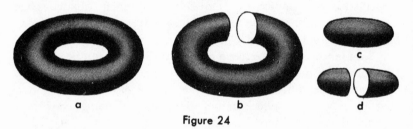

Figure 24

What about the hole in a doughnut? Is the hole in a doughnut inside or outside? In topology, we say the hole is outside, not inside. Actually, the hole in the doughnut has only a small part in topological classification definitions.

Next let's look at the edges of topological figures. A sheet of paper or a round card is said to have two surfaces and one edge. An inner tube has two surfaces but no edge. An open cylinder has two edges and two surfaces. One way to classify objects is to count the cross-cuts that can be made on a surface without dividing it into more than one piece. A cross-cut may be thought of as a cut with a pair of scissors that begins and ends on an edge.

If we make a cut from one edge of a square sheet of paper to another, we see that we divide it into two distinct parts. If we do this with a cardboard tube, we reduce it to a surface equivalent to a square. Of course, if we cut the cylinder by a line parallel to both of its edges, we get two cylinders. The square is said to be a simple surface; the cylinder, a *singly connected surface*. A punctured sphere (balloon) could be stretched and flattened into a sheet if the hole would stretch enough. So a sphere is also a singly connected surface.

Each of the three-dimensional objects in Figure 25 is *triply connected* because two cuts are needed for each to be changed to a simple closed surface, and then a third cut or puncture is needed to change the closed surface to a simple plane curve.

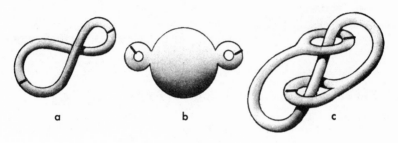

Figure 25

EXERCISE SET 7
Three-Dimensional Surfaces

1. How many cuts are needed to change these solids to simple closed surfaces?

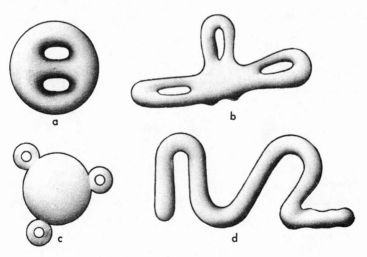

2. Classify these surfaces as singly. doubly, or triply connected surfaces or as simple surfaces.

a. a football   e. an inner tube
b. a garden hose   f. a paper plate
c. a coat   g. a paper cup with
d. a pullover sweater   no handles

3. Because of differences in connectivity, some puzzles that cannot be solved on a simple closed surface like a plane or a sphere can be solved on a more complex surface like a torus (the technical name for a doughnut-like surface).

You should have found it impossible to do Problem 1 of Exercise Set 4 in one plane. See if you can solve this problem on a doughnut.

A Famous Mathematician's Unusual Bottle

A three-dimensional object similar to a one-sided Moebius strip is the Klein bottle, invented in 1882 by the great German mathematician, Felix Klein. The easiest way to visualize this bottle is to imagine that an inner tube is cut and straightened out like a cylinder. One end is then stretched out to make a base and the other end narrowed like the neck of a bottle. Next, the

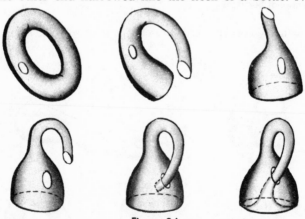

Figure 26

narrow end is twisted over and thrust through the valve-stem hole in the side of the tube. Finally, this end is flared out and joined with the open end at the base. This may be called a "punctured" Klein bottle, the hole in the tube being the puncture in the bottle. In topology, we usually suppose that no hole actually exists, so that the one-sided surface passes through itself. This, of course, is actually impossible to do with an inner tube, but in topology we make free use of such queer possibilities. A Klein bottle may be thought of as a pair of Moebius bands with the edges glued together.

These properties of Moebius bands and Klein bottles have been summarized in a pair of limericks:

> A mathematician confided
> That a Moebius band is one-sided.
>   And you'll get quite a laugh
>   If you cut one in half,
> For it stays in one piece when
>   divided.

> A mathematician named Klein
> Thought the Moebius band was
>   divine.
>   Said he, "If you glue
>   The edges of two,
> You'll get a weird bottle like
>   mine."

Euler's Formula Looks at the Third Dimension

Euler's network formula, $V - A + R = 2$, can be applied to certain three-dimensional figures called *polyhedrons*. A polyhedron is a solid made up of parts of plane surfaces called *faces* of the polyhedron. A brick is one example of a polyhedron. In a *regular polyhedron*, such as a cube, the faces are geometric figures with all sides and all angles equal, all faces have the same shape and size (are congruent), and the angles at which the faces meet can be made to coincide.

There are five and only five regular polyhedrons. We have already mentioned the cube as one example of a regular polyhedron. A polyhedron with six faces is called a *hexahedron*. Since a cube is a regular polyhedron with six faces, it is called a *regular* hexahedron. The other polyhedrons that can be regular are those with four faces *(tetrahedron)*, eight faces *(octahedron)*, twelve faces *(dodecahedron)*, and twenty faces *(icosahedron)*. The regular polyhedrons are pictured in Figure 27. Find models of these polyhedrons. You can make models yourself out of cardboard, using the patterns of Figure 28.

When applying Euler's formula to polyhedrons, we often change the symbol A to E to represent the number of edges of the polyhedron, and change R to F to represent the number of faces. Do you see that a polyhedron is really a three-dimensional network?

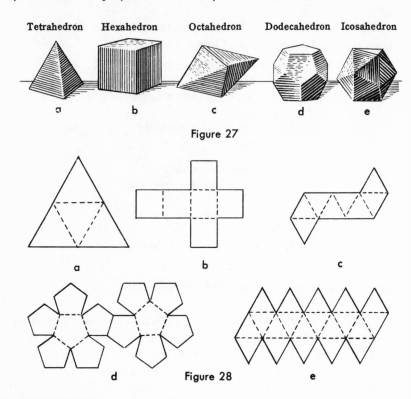

Figure 27

Figure 28

EXERCISE SET 8

Polyhedrons and Euler's Formula

Copy this table and fill in the blanks.

Name	Number of Faces (F)	Number of Edges (E)	Number of Vertices (V)	V+F	E+2
1. Tetrahedron					
2. Hexahedron					
3. Octahedron					
4. Dodecahedron					
5. Icosahedron					

Stunts and Fun
with Topology

Topology is such a curious kind of mathematics that it can very easily be used to mystify people who are not acquainted with it. We saw one example of this with the Moebius strip. There are many other odd and unusual effects that can be achieved through various applications of topology, and you may want to present these at parties and other gatherings for the delight and/or mystification of your friends. Also, topology has many interesting problems and puzzles that are fun to try to solve at gatherings. Some of these are given in the following pages. Why not try them? If you need help, you will find the solutions at the back of this book.

Knots and Topology

One application of topology is the study of knots in a string. If a knot is loosely tied, it is possible to work it along the string toward the end of the string. Suppose we tie two knots in a string, as in Figure 29.

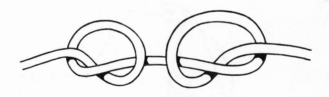

Figure 29

Now work the knots toward each other. These knots are the opposite of each other. Yet they will not untie each other when brought together. Instead, one passes through the other and out the other side, leaving both knots unchanged. Experiments show that this is always true with these knots, but no one has been able to prove it.

A false knot known as the Chefalo Knot is an example of knots which are used by magicians. It begins as a square knot as in

Figure 30a. Then one end is woven in and out as shown in Figure 30b by arrows. When the ends are pulled, the knot disappears.

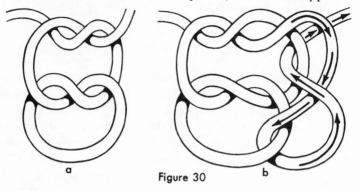

Figure 30

Buttonholing a Friend

Tie a loop of string to a pencil or a short stick. Be sure the loop is shorter than the pencil. Attach the pencil to the buttonhole of a friend's jacket without untying the loop, as shown in Figure 31b. Pull the loop tight, as shown in Figure 31c. Ask your friend to remove the pencil without untying or cutting the loop (or his buttonhole!). If he doesn't see you put it on, he will have a hard time removing it.

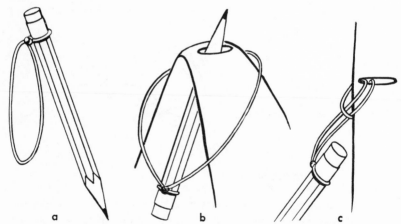

Figure 31

A variation of this puzzle is to loop a string through a pair of scissors and then tie the ends of the string to a large button, as shown in Figure 32. The button must be larger than the opening in the handle of the scissors. The problem is to remove the button from the pair of scissors without untying or cutting the string.

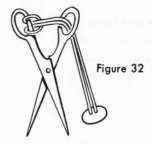

Figure 32

De-vesting Yourself

The object of this stunt is to remove your vest without removing your coat. Put on a vest and a coat. If the vest is large, the stunt is easier to perform. The coat may be unbuttoned, but you are not permitted to let your arms slip out of your coat sleeves.

The Ring Puzzle

The three rings pictured below have a strange topological relation. Remove any one ring, and the other two will be found to be free, too. Thus, no two rings are joined, but the three put together are.

Figure 33

Stringing Along

Tie a piece of string to each of your wrists. Tie a second piece of string to each of the wrists of a partner in such a way that the second string loops the first. The object of this stunt is to separate yourself from your partner without cutting the string, untying the knots, or taking the string off your wrists. This can be done!

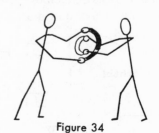

Figure 34

227

Buttons and Beads

To make this puzzle you need cardboard, string, two buttons,

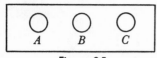

and two beads. Cut a rectangular piece of cardboard about 1 inch by 6 inches. Cut three small, evenly spaced holes, as in Figure 35.

Figure 35

String two large beads on the string. Thread one end of the string through hole *A* and attach a button larger than the hole. In the same direction, thread the other end of the string through hole *C* and attach a button as in Figure 36*a*.

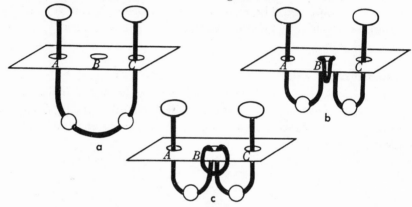

Figure 36

The string is then looped through hole *B*, as in Figure 36*b*. To loop it back under itself, as in Figure 36*c*, the loop is first threaded up in hole *A* and over the button and then likewise in hole *C*. The puzzle is now ready for someone to try to undo the loop and get the beads together.

The Swiss School Problem

Four Swiss schoolboys live at homes *A*, *B*, *C*, *D*. They go to the same school and must enter doors *A*, *B*, *C*, *D*. Boy *A* lives in

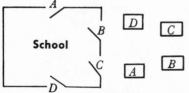

home *A* and goes to door *A*, *B* goes from home *B* to door *B*, and so on. How can they ski to school without having their tracks cross?

Figure 37

The Divided Farm

A wealthy farmer with four sons owned a three-section farm having the shape shown in Figure 38. When he died, his will said that the farm should be divided so that each son got one-fourth

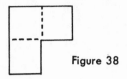

Figure 38

of his farm. He also stated that each farm must have the same shape as the original farm. How was the farm divided?

Pennies on the Square

Draw a 6-by-6 square checkerboard. Place six pennies on this board so that no penny is in line with another penny horizontally, vertically, or diagonally. No square may have more than one penny.

The Street Sweeper's Route

A street sweeper must sweep each of the streets shown in Figure 39. To cover the area, he must drive some blocks twice. How should he plan his route so that he can cover it by traveling the least number of blocks possible? He may start at any corner.

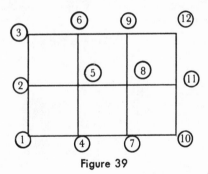

Figure 39

The Paper and String Puzzle

Take a piece of stiff paper and cut it so it measures 6 inches long by 3 inches wide. Now cut two parallel slits half an inch apart in the center of the paper, as shown in Figure 40. Each slit is 3 inches long. Half an inch above the slits, cut a circular hole having a diameter of ¾ inch.

Pass a piece of string about 12 inches long behind the slits and then down through the hole, as shown in the drawing, and tie a large button to each end of the string. Be sure that each of the buttons is too large to pass through the ¾-inch hole in the paper.

Now ask one of your friends to remove the string and buttons from the paper without tearing the paper or taking off either of the buttons. There is little chance of his succeeding unless you show him how.

Figure 40

The Paper Boot Puzzle

The boot puzzle consists of three pieces, all of which are cut out of stiff paper. One piece is shaped like a pair of boots joined together at the top, as in Figure 41a. The remaining pieces are shaped as shown in Figure 41b and c.

To assemble the puzzle, fold the large rectangular piece as in Figure 41b and slip the smaller piece over one of its arms, as in Figure 41d. Then hang the boots over part of the same arm, as in Figure 41e. Pull the small piece to the right and over the end of the arm at A. Then unfold the large piece, and the puzzle will be assembled as in Figure 41f.

The problem is to remove the boots without tearing the paper.

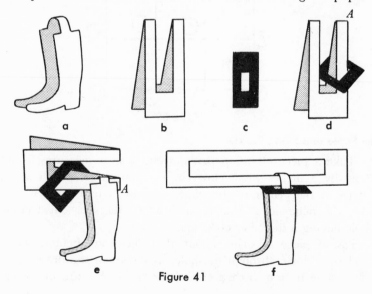

Figure 41

EXERCISE SET 9

Topology Review Test

Take this test to see how much you remember about the ideas you have come across in this part of the book.

1. Which figure in each set of four does not belong with the rest?

2. Which of the following networks can be traveled in one journey?

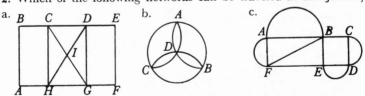

3. Which of the following are invariant under a topological transformation?

      a. The number of regions in the figure.

      b. The value of $(V+R-A)$ in a network.

      c. The length of a line.

d. Continuity.

e. The shape of the figure.

4. What is the minimum number of trips needed to trace this network?

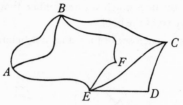

5. Name all vertices where you can start to trace the above network in the minimum number of trips.

6. The following diagram shows a fleet of boats which are tied up near a dock. The boats are connected by gangplanks as indicated. Show by use of a network whether or not it is possible to make one trip and cross all of the gangplanks only once. If not, what is the minimum number of trips?

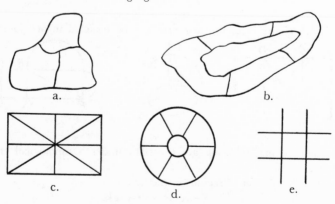

7. Show that Euler's formula for vertices, arcs, and regions is satisfied for each of the following figures.

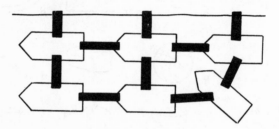

232

Demonstration Models for Topology

In addition to the models, drawings, and stunts described in this section, there are other models to demonstrate the ideas of topology. These will dress up and add interest to your science exhibit, class report, or assembly program. Here are some suggested models.

1. Make drawings of a geometric figure on several rubber sheets. Show how the figure changes when the sheets are stretched in various ways.

2. Cut 12 lengths of soda straws each about 2 inches long. Thread elastic thread through these straws to form a cube. Show how this cube can be distorted or flattened or changed into a network.

3. Draw a network or map on several balloons. Show how this network or map remains equivalent with the balloons blown up to different sizes.

4. Stretch a rubber sheet or balloon over a geometric solid. Show how the geometric solid and a sphere have the same characteristics.

5. Use modeling clay to form models of objects to show different types of topological surfaces.

6. Make a Klein bottle out of rubber or knit one out of yarn.

7. Make large Moebius strips out of adding machine tape. Cut in different ways to show the many possible results.

8. Make diagrams of complex maps to demonstrate the four-color problem.

We have presented some of the basic concepts and interesting applications of topology. But this field of study is not concerned only with tricks, puzzles, and fascinating problems. A technical study of topology is often a part of advanced college work in mathematics, and involves many ideas far beyond the scope of this book. These concepts are developed in much the same way that theorems are developed in high-school geometry. Mathematicians are applying topological concepts to many practical problems, and topology is becoming a valuable tool in our complex space age.

Extending Your Knowledge

If you would like to learn more about topology, the references below will tell you many more unusual things about this subject.

COURANT, RICHARD and ROBBINS, HERBERT, "Topology," in James Newman's *The World of Mathematics*. Simon and Schuster, 1956

EULER, LEONHARD, "The Seven Bridges of Koenigsberg," in James Newman's *The World of Mathematics*. Simon and Schuster, 1956

GAMOW, GEORGE, *One, Two, Three, Infinity*. Viking Press, 1947

KASNER, EDWARD and NEWMAN, JAMES, *Mathematics and the Imagination*. Simon and Schuster, 1940

MESERVE, BRUCE E., "Topology for Secondary Schools," in *The Mathematics Teacher*, November, 1953

TUCKER, A. W. and BAILEY, H. S., "Topology," in *Scientific American*, January, 1950

If you like science fiction, you will like these stories that are based on topology.

DEUTSCH, A. S., "A Subway Named Moebius," in Clifton Fadiman's *Fantasia Mathematica*. Simon and Schuster, 1958

GARDNER, MARTIN, "No Sided Professor," in *Esquire*, January, 1947

"The Island of Five Colors," in Clifton Fadiman's *Fantasia Mathematica*. Simon and Schuster, 1958

UPSON, WILLIAM HAZLETT, "A. Botts and the Moebius Strip," in *Saturday Evening Post*, December 27, 1945

PART VI

Fun
WITH MATHEMATICS

Mathematical Fun

Surprises and Stunts

There is nothing quite as much fun as a surprise, particularly if the surprise is such that your friends think you have special powers. There are many problems in this part of the book that will give you special abilities. You can tell people what change they have in their pocket, what telephone number they have, their address, the number of people in their family, and even their age. Yes, you can certainly make others think that you have supernatural powers.

The secret of these powers lies not in the supernatural, however, but in the fascinating world of mathematics. Every trick or problem in this section has a mathematical basis. As a result, you can not only see why the problems behave the way they do, but you will also be able to develop more problems of a similar type. In fact, there is no limit to the mathematical tricks you can develop for yourself.

There are many interesting ways you can use these materials, too. One of the most obvious is just to amaze your family or neighborhood friends, or even yourself. You will also find a good audience in your mathematics classroom because each problem illustrates applications of some branch of mathematics. A few at a time are especially good to use at meetings of a mathematics club. You can even use them in school assembly skits, if you dress them up as mind-reading or magic acts.

So, with these suggestions, have fun in studying each problem. Try them out on as many people as possible. And above all, see what new problems you can invent by using your knowledge of mathematics.

Bending the Thoughts of Others to Your Will

Take a Number

Have you ever had the experience of having a friend tell you what number you are thinking about? Perhaps you were asked to take a number and do certain things to it, like adding, subtracting, multiplying, and dividing by some numbers. Then, as if by magic, your friend suddenly told you what answer you had, even though he didn't know what number you had picked at the beginning.

Let's try a trick like this. Just follow the steps given below.

Step 1. Take a number.
Step 2. Add 3.
Step 3. Multiply by 2.
Step 4. Subtract 4.
Step 5. Divide by 2.
Step 6. Subtract the number you started with.

If you have done the computations correctly, your answer is 1.

If you think this is just a coincidence, try another number. Your answer, after following the six steps, will again be 1. No matter what number you choose, the answer will always be 1.

Although this perhaps sounds like magic, it is easily explained by mathematics. Let's give an explanation first with actual objects, such as marbles.

You can start with any number, so let's show this by a bag of marbles, . No one can see the number of marbles in your particular bag. Now let's follow the steps again.

Step 1. Take a number.	Show this as a bag:	
Step 2. Add 3.	The bag plus 3 marbles:	
Step 3. Multiply by 2.	Two bags plus 6 marbles:	
Step 4. Subtract 4.	Two bags plus 2 marbles:	
Step 5. Divide by 2.	One bag plus 1 marble:	
Step 6. Subtract the number you started with.	1 marble is left:	•

In place of the picture of a bag of marbles, we can use the letter N to represent any number. A symbol, such as N, that holds a place for a number, is sometimes called a *variable*. The steps in the trick now look like this:

Step 1. Take a number.	Instead of	use N.
Step 2. Add 3.	Instead of	use $N+3$.
Step 3. Multiply by 2.	Instead of	use $2N+6$.
Step 4. Subtract 4.	Instead of	use $2N+2$.
Step 5. Divide by 2.	Instead of	use $N+1$.
Step 6. Subtract the number you started with.	Instead of •	use 1.

We can now use number and operation symbols to express each step that is done in this trick. The steps can be shown like this:

Step 1. Take a number. N

Step 2. Add 3. $N + 3$

Step 3. Multiply by 2. $2(N + 3)$

Step 4. Subtract 4. $2(N + 3) - 4$

Step 5. Divide by 2. $\dfrac{2(N + 3) - 4}{2}$

Step 6. Subtract the number $\dfrac{2(N + 3) - 4}{2} - N$
you started with.

The expression $\dfrac{2(N + 3) - 4}{2} - N$ represents a combination of all the steps taken to get the final result. We know that this expression must equal 1 for all values of N. Let's examine the algebraic steps that can be taken in a logical order to make the expression simpler without changing its value.

The simplification of $\dfrac{2(N + 3) - 4}{2} - N$ starts with $N + 3$. The symbols say, "Double $(N + 3)$," which becomes $2N + 6$, and the expression can be rewritten as $\dfrac{2N + 6 - 4}{2} - N$.

Next, the symbols say, "Subtract 4 from $2N + 6$." This gives the result $2N + 2$, and the expression becomes $\dfrac{2N + 2}{2} - N$.

The expression $\dfrac{2N + 2}{2}$ means, "Divide $2N + 2$ by 2." This gives $N + 1$ and, therefore, the expression simplifies to $N + 1 - N$.

This now says, "Subtract N from $N + 1$." The answer is 1.

Step by step, we have taken a complicated expression, $\dfrac{2(N + 3) - 4}{2} - N$, and changed it to a much simpler form, 1.

This is what is called an *identity* in algebra, and it is written as $\dfrac{2(N + 3) - 4}{2} - N \equiv 1$. The right-hand side equals the left-hand side for *all* values of N.

We can now see why this trick will always give the answer 1. There is no mind reading involved; the trick uses mathematical principles. After studying examples like this, you will be able to make original problems to amaze your friends.

In explaining this trick, we let a letter, N, play the role of a "placeholder" which can be filled by any number. Thus N is called a *variable*, and the set of numbers that can serve as replacements for the variable is called the *replacement set*. We then apply operational principles of arithmetic to expressions containing variables. Such an examination of problems involving our number system is an important part of the branch of mathematics called algebra.

5 Can Be a Lucky Number

We can show the versatility of the preceding trick by doing a variation that always gives 5 for the answer. We will do a particular example first and then show the generalization for any number, N.

Start with any number, such as	15
Add the next larger number.	$15 + 16 = 31$
Add 9 to this sum.	$31 + 9 = 40$
Divide the result by 2.	$\dfrac{40}{2} = 20$
Subtract the original number.	$20 - 15 = 5$

The answer is 5, and it is always so no matter what number was chosen at first.

This last statement needs proof, and a good way to prove it is to use the variable N to hold a place for any number and to show that, if you follow the steps in succession, the last step will give an answer of 5 — always!

Here is the proof:

Start with a number.	N
Add the next larger number, $(N+1)$.	$N+(N+1)=2N+1$
Add 9 to this sum.	$2N+1+9=2N+10$
Divide the result by 2.	$\dfrac{2N+10}{2}=N+5$
Subtract the original number.	$N+5-N=5$

Thus, the answer is 5, and the proof is complete.

Again, what first appeared to be a baffling trick becomes a simple exercise in algebra.

EXERCISE SET 1
"Name-the-Answer" Tricks

1. Suppose you wish to cast a spell upon a friend such that, no matter what number he chooses, the result of his following a prescribed set of instructions will always be 13.

First, you may ask him to choose any number he may wish. Then you direct him through a set of operations carefully designed to force him to eliminate the number which he has chosen and to put in the number 13.

Test the following procedure with any number:

> Choose a number.
> Add 11.
> Multiply by 6.
> Subtract 3.
> Divide by 3.
> Subtract 6 less than the original number.
> Subtract 1 more than the original number.
> Divide by 2.
> The answer is 13.

The same result could have been obtained if you had simply asked your friend to choose a number, to subtract the number chosen from itself, and then to add 13. The complicated set of instructions given really served only as a device to complicate the trick.

See if you can prove this procedure by methods of algebra.

2. Take any number; multiply it by 2; add 16 to that result; divide by 2; and then subtract the original number. What is the answer? Prove that you would obtain this answer no matter what starting number you used.

Pick Your Answer...
Then Find the Problem

Starting with N

We have used algebra to explain number tricks, but we can also use the same methods to develop new tricks and problems. In the following examples, we start with any number, N, perform a series of numerical operations, and obtain an expression that is very simple to interpret.

	Example 1	Example 2	Example 3
Step 1.	Start with N.	Start with N.	Start with N.
Step 2.	Add 2.	Subtract 2.	Multiply by 2.
	$N+2$	$N-2$	$2N$
Step 3.	Double.	Triple.	Add 10.
	$2N+4$	$3N-6$	$2N+10$
Step 4.	Subtract 2.	Add 6.	Subtract 4.
	$2N+2$	$3N$	$2N+6$
Step 5.	Divide by 2.	Divide by 3.	Divide by 2.
	$N+1$	N	$N+3$
Step 6.	Subtract N.	Subtract N.	Subtract N.
	1	0	3

Thus we have developed three tricks that will always give a certain number as an answer no matter what starting number is used. Example 1 will always give the answer 1, Example 2 will always give the answer 0, and Example 3 will always give the answer 3.

In each example, we can obtain another interesting type of problem by using only the first five steps. If we have the answer to Step 5 in each example, we can quickly determine the original numbers.

The starting number, N, for Example 1 would be 1 less than the answer to Step 5.

The starting number, N, for Example 2 is the same as the answer to Step 5.

The starting number, N, for Example 3 is 3 less than the answer to Step 5.

You can develop tricks in which the final answer will tell you a starting number.

You can apply this principle to a variety of interesting problems where the numbers which people select may be their age, their address, the change they may have in their pocket, or anything else that they may wish to think of. Nobody will be able to keep secrets from you if they follow your directions and give correct answers to your questions.

Who Should Pay the Check?

Ask your friend to count all of the loose change he has in his pocket. Tell him that you will be able to find how much change he has if he just shows you the final answer to these steps. Ask him to start with the change, say 64 cents.

Multiply the amount by 2.	$2(64) = 128$
Add 3 to the product.	$128 + 3 = 131$
Multiply the sum by 5.	$5(131) = 655$
Subtract 6.	$655 - 6 = 649$

Ask him to show you the final answer, 649, and you can tell him immediately that the change is 64 cents. Just cross out the ones digit, 64$\not9$, and the remaining digits express the change.

Here is the algebraic proof:

Start with the change. C
Multiply by 2. $2C$
Add 3. $2C + 3$
Multiply by 5. $10C + 15$
Subtract 6. $10C + 9$

This kind of problem is good for finding any number — age, address, telephone number. You can also find more than one number at a time. Notice how this can be done in the next two examples.

Old Enough for a Driver's License?

This trick will permit you to determine a person's age and, at the same time, tell the amount of change he is carrying.

Ask a friend to write his age on a piece of paper. Suppose his age is 16.

Multiply by 4. $4(16) = 64$
Add 10. $64 + 10 = 74$
Multiply by 25. $25(74) = 1850$
Subtract the number of days in the
  year (365; not leap year). $1850 - 365 = 1485$
Add the change less than one dollar
  in his pocket, say 38 cents. $1485 + 38 = 1523$
Add 115. $1523 + 115 = 1638$

When your friend gives you this answer, you can immediately tell him his age and the change he has because: the first two digits in the answer are 16, his age, and the last two are the change, 38 cents.

This trick can also be explained by using algebra:

Start with age.	A
Multiply by 4.	$4A$
Add 10.	$4A + 10$
Multiply by 25.	$25(4A + 10) = 100A + 250$
Subtract the number of days in the year, 365.	$100A + 250 - 365 = 100A - 115$
Add the change less than one dollar, C, in his pocket.	$100A - 115 + C$
Add 115.	$100A - 115 + C + 115 = 100A + C$

$100A$ places the ones digit of A in the hundreds place of the final answer. C is then in the ones place or the ones and tens places.

If the change is more than a dollar, the numeral in the hundreds place will be determined partly by the change and partly by the age; hence you will not be able to separate the two.

How Many Candles for the Cake?

Here is a trick that will enable you to name the date of a friend's birthday. Assign numbers to each month in order so that January is 1, February is 2, March is 3, and so on.

Then tell your friend to make the following computations. If he tells you the answer at the end, you will be able to tell him the month and day of his birthday.

Suppose his birthday is April 25.

Multiply the number of the month by 5.	$5(4) = 20$
Add 7.	$20 + 7 = 27$
Multiply by 4.	$4(27) = 108$
Add 13.	$108 + 13 = 121$
Multiply by 5.	$5(121) = 605$
Add the day of the month.	$605 + 25 = 630$

Ask your friend for the answer. You will mentally subtract 205 from 630 and obtain 425.

The 4 stands for April, the 25 is the day of the month. You can say to your friend, "Your birthday is on April 25."

Algebra again supplies the explanation.

Let M represent the number of the month and D the day of the month.

Multiply the number of the month by 5.	$5M$
Add 7.	$5M + 7$
Multiply by 4.	$4(5M + 7) = 20M + 28$
Add 13.	$20M + 28 + 13 = 20M + 41$
Multiply by 5.	$5(20M + 41) = 100M + 205$
Add the day of the month, D.	$100M + 205 + D = (100M + D) + 205$

If you subtract 205, the $100M + D$ exhibits the number of the day in the tens and units places and the month in the hundreds and thousands places.

EXERCISE SET 2
More Number Tricks

1. Perform the following operations:

> Multiply your age by 2.
> Add 10.
> Multiply by 5.
> Add the number of people in your family.
> Subtract 50.

Your age appears in the hundreds and tens places of the resulting numeral, and the number in your family is shown in the ones place. Use algebra to prove this trick. Will this trick always work?

2. Double your house number; add the number of days in a week; multiply by 50; add your age; subtract 365, the number of days in a year; add 15. The tens and units digits of the resulting numeral give your age and the other digits are the house number. Use algebra to explain this trick.

3. Devise a trick that will instruct a person to select a number and perform a series of computations, but will always give 4 as a result, no matter what original number was selected.

4. Devise a trick that will instruct a person to select a number and perform a series of computations, but will always produce a number 1 less than the original number.

Magic in the Number System

Algebra Opens the Way

We have used algebra to explain and discover some number tricks, but this is only one minor use of this important branch of mathematics. Through a study of algebra, we can examine the logic, organization, and development of our system of numbers. Such an exploration of our number system can lead to many interesting facts that will seem quite mysterious to many people. Thus, you can again assume the role of number magician.

Odd or Even?

Every performer needs his stage props, and in this instance you are no exception. Several pennies are needed as props for this trick. If the financial strain is too great to raise any amount of money, you can, of course, let people imagine that they have pennies. Let's assume that you can muster up the pennies.

Ask a friend to place an odd number of pennies in one hand and an even number of pennies in his other hand. Ask him to double the number in his left hand and triple the number in his right hand. Then have him give you the sum of these two amounts, and you will be able to tell him which hand has the odd number in it and which has the even number in it.

Here is how you can tell:

If the total is odd, he will have an odd number in his right hand.

If the total is even, he will have the odd number in his left hand.

Suppose the left hand has five pennies (odd) and the right hand has eight pennies (even).

Double what is in the left hand. $2(5) = 10$

Triple what is in the right hand. $3(8) = 24$

Find the total. $\overline{34}$

Since 34 is even, the odd number, according to the rule, must be in the left hand.

Let's try to discover an explanation for this trick. We know that any even number is divisible by 2. This idea can be expressed algebraically by saying that an even number is one that can be written in the form $2K$, where K can be any suitable whole number. For example, 18 is even because it can be put into the form $2(9)$, if we let K equal 9.

An odd number is one which can be expressed in the form $2K + 1$. The number 23 is odd because it can be put into the form $2(11) + 1$, if we let K equal 11.

With this in mind let's consider the two possible situations that can exist. We will use K_1 and K_2 to represent any two whole numbers.

Case 1. Odd number in the left hand. $2K_1 + 1$

Even number in the right hand. $2K_2$

Double the odd number. $2(2K_1 + 1)$, which is even because it can be divided by 2 with no remainder resulting.

Triple the even number. $3(2K_2)$, which is also even.

Add these two even numbers. $2(2K_1 + 1) + 3(2K_2) =$ $4K_1 + 2 + 6K_2 = 2(2K_1 + 3K_2 + 1)$, which is even because it is divisible by 2.

Case 2. Even number in the
      left hand. $2K_2$

Odd number in the
      right hand. $2K_1 + 1$

Double the even
      number. $2(2K_2)$, which is evenly divisible by 2 and hence is even.

Triple the odd
      number. $3(2K_1 + 1) = 6K_1 + 3$, which is an odd number, for it is not evenly divisible by 2.

Add the even number
      and the odd number. $2(2K_2) + 6K_1 + 3 = 4K_2 + 6K_1 + 3$, which is odd because it is not evenly divisible by 2.

Next-Door Neighbors in the Number System

Everybody knows that any whole number in our number system is equal to the number that comes before it, increased by one. However, few people are aware that this fact can be useful in performing feats of number magic. Therefore, the following trick may be worthy of a listing in your catalog of sorcery.

Ask a friend to choose a number, perhaps 386.

Then ask him to add to this number the next four consecutive whole numbers.

$$
\begin{array}{r}
386 \\
387 \\
388 \\
389 \\
\underline{390} \\
1940
\end{array}
$$

If he will tell you this sum, 1940, you can find the number he started with by dividing the sum given, 1940, by 5 and then subtracting 2.

$$\frac{1940}{5} = 388$$

$$388 - 2 = 386$$

You will always get the number you chose at the start.

Algebra can be used to write a proof for this problem.

Use the variable N to represent any number.

Add N and the next four consecutive whole numbers:

$$N + (N+1) + (N+2) + (N+3) + (N+4) = 5N + 10$$

Divide by 5:

$$\frac{5N + 10}{5} = N + 2$$

Subtract 2:

$$N + 2 - 2 = N$$

You get the number N which you chose at the start.

A Place-Value Puzzler

Here is a trick that requires a knowledge of the place-value concept used in our numeration system.

Take any four-place number, such as 6245.

Write the thousands digit.	6
Write the thousands and hundreds digits.	62
Write the thousands, hundreds, and tens digits.	624
Add these numbers.	$6 + 62 + 624 = 692$
Multiply by 9.	$9 \times 692 = 6228$
Find the sum of the digits in the original number.	$6 + 2 + 4 + 5 = 17$
Add this sum to the previous result.	$6228 + 17 = 6245$

The answer will always be the same as the four-place number originally chosen.

The explanation of this trick is closely related to the place-value principle of our number system.

Any four-place number may be written as

$$a_3(1000) + a_2(100) + a_1(10) + a_0,$$

where a_2, a_1 and a_0 hold a place for any digit, and a_3 holds a place for any digit except 0.

Add the three numbers as specified:

$$[a_3] + [a_3(10) + a_2] + [a_3(100) + a_2(10) + a_1]$$
$$= a_3(111) + a_2(11) + a_1(1).$$

Multiply by 9:

$$a_3(999) + a_2(99) + a_1(9).$$

Add the sum of the digits to this result:
$$[a_3(999) + a_2(99) + a_1(9)] + [a_3 + a_2 + a_1 + a_0].$$
But this gives the original number,
$$a_3(1000) + a_2(100) + a_1(10) + a_0.$$

A Square Can Be a Welcome Friend

Here is a method for determining an unknown number by applying knowledge about the properties of squares and the process of raising a number to the second power.

Take a number, such as 3.

Square it. $\qquad$ $3^2 = 9$

Square the next larger number. $\qquad$ $(3 + 1)^2 = (4)^2 = 16$

Find the difference. $\qquad$ $16 - 9 = 7.$

Subtract 1 from 7, and then divide by 2 to find the original number. $\qquad$ 3.

An algebraic proof unveils the mystery of this trick:

Take any number, N.

Square it. $\qquad$ N^2

Square the next larger number. $\qquad$ $(N + 1)^2$

Find the difference. $\qquad$ $(N + 1)^2 - N^2$
$$= N^2 + 2N + 1 - N^2$$
$$= 2N + 1$$

Subtract 1. $\qquad$ $2N + 1 - 1$
$$= 2N$$

Divide by 2. $\qquad$ $\dfrac{2N}{2} = N$

Camouflage Adds Spice

Very often we can make our number magic seem much more mysterious by inserting into our procedure some restriction or limitation which may really have no effect on the results.

Let's try another exercise involving the process of squaring and introduce the restriction that our beginning number be less than 25.

Choose any number less than 25, such as 15.

Add it to 25. $25 + 15 = 40$

Subtract it from 25. $25 - 15 = 10$

Multiply each of these
numbers by itself. $40 \times 40 = 1600$
$10 \times 10 = 100$

Subtract the smaller
from the larger. $1600 - 100 = 1500$

The answer is always
100 times the
original number. $1500 = 15(100)$

The following algebraic proof explains this trick:

Take a number. N (N is less than 25)

Add it to 25. $25 + N$

Subtract it from 25. $25 - N$ (Since N is less than 25, then $25 - N$ is a positive quantity.)

Multiply each of these
quantities by itself. $(25 + N)^2 = 625 + 50N + N^2$
$(25 - N)^2 = 625 - 50N + N^2$

Subtract the smaller
from the larger. $(625 + 50N + N^2) - (625 - 50N + N^2)$
$= 100N$

This shows that the answer is always 100 times the original number.

Was the restriction that N be less than 25 really necessary? Test the result of this trick with several numbers greater than 25.

A Quotient and Difference Puzzler

This exercise should be sufficiently startling to make a hero of you in any group. Two simple computations made by an assistant empower you to make known two secret numbers.

Let your friend choose any two numbers, say 12 and 4.

Divide the larger by the smaller. $\dfrac{12}{4} = 3$

Subtract the smaller from the larger. $12 - 4 = 8$

If your friend tells you this quotient and this difference, you can find the smaller of the numbers chosen by dividing one less than the quotient into the difference: $\frac{8}{3-1} = 4$.

You can find the larger number by multiplying the smaller number by the quotient: $(4)3 = 12$.

Here is the explanation:

Choose two numbers.	A and B. (Assume A is less than B.)
Divide the larger by the smaller.	$\frac{B}{A} = Q$ (the quotient)
Subtract the smaller from the larger.	$B - A = D$ (the difference)
Then, from the fact that $\frac{B}{A} = Q$,	$B = A(Q)$
From the fact that $B - A = D$,	$B = A + D$
But $B = B$; therefore,	$A(Q) = A + D$
Subtract A from both sides of the equation.	$A(Q) - A = D$
Factor the left side.	$A(Q - 1) = D$
Divide both sides by $Q - 1$.	$A = \frac{D}{Q - 1}$

And A is the smaller of the two numbers chosen.

Be a Computing Wizard

You may never find a practical application for this trick, but you can amaze friends by appearing to have a tremendous ability to compute rapidly. Have two people each give you a four-digit number. Suppose one number is 1776 and the other is 1959.

Have one person multiply the two numbers:

```
        1 7 7 6
      × 1 9 5 9
      ─────────
        1 5 9 8 4
        8 8 8 0
      1 5 9 8 4
      1 7 7 6
      ─────────
      3,4 7 9,1 8 4
```

Mentally subtract 1776 from 10,000 and 1 from 1959, obtaining 8224 and 1958. Ask the other person to multiply 8224 and 1958:

$$
\begin{array}{r}
8\,2\,2\,4 \\
\times\ 1\,9\,5\,8 \\
\hline
6\,5\,7\,9\,2\quad \\
4\,1\,1\,2\,0\quad\ \ \\
7\,4\,0\,1\,6\quad\ \ \ \ \\
8\,2\,2\,4\quad\ \ \ \ \ \ \\
\hline
1\,6{,}1\,0\,2{,}5\,9\,2
\end{array}
$$

It might be a good idea to have the two people check each other's multiplication. Then have them add the two answers; and while they are doing this, you can write the final answer on a piece of paper. They will be amazed to find that what you have written agrees with the sum of their two answers.

You write: 19,581,776.

They add:
$$
\begin{array}{r}
3{,}4\,7\,9{,}1\,8\,4 \\
+\ 1\,6{,}1\,0\,2{,}5\,9\,2 \\
\hline
1\,9{,}5\,8\,1{,}7\,7\,6
\end{array}
$$

And you point out, quite proudly, that these agree.

The proof by algebra follows:

Each person gives you a four-digit number.	M and N
The first person multiplies them.	MN
The second person is given two other numbers to multiply. They must be	$10{,}000 - M$ and $N - 1$
Multiplying them gives	$(10{,}000 - M)(N - 1) =$ $10{,}000N - 10{,}000 - MN + M =$ $10{,}000(N - 1) - MN + M$
You ask the two friends to add their answers:	$(MN) + [10{,}000(N - 1) - MN + M] =$ $10{,}000(N - 1) + M$

This shows why you can write down the answer immediately. Multiplying a number by 10,000 sets the digits over four places.

The first four places in the answer are the digits $N-1$ and the next four places are the digits in M.

In the example, M was 1776 and N was 1959. Therefore, you were able to write the answer as 19,581,776. The first four digits were $N-1$, or 1958, and the last four digits were M, or 1776.

EXERCISE SET 3
Puzzlers from Our Number System

1. Take any number; add 4; multiply by 3; subtract 9; multiply by 3; add the original number; scratch out the units digit of the resulting numeral. The number obtained is the original number. Prove this by using algebra.

2. Work out a procedure to determine an unknown number, N, when the sum of N and the next six consecutive whole numbers are known.

3. Work out a procedure to determine an unknown number, N, when the sum of N and the next X consecutive whole numbers are known.

4. Consider two numbers. A and B. (Assume A is less than B.)

Divide the larger by the smaller. $\dfrac{B}{A} = Q$ (the quotient)

Subtract the smaller from the larger. $B - A = D$ (the difference)

Write a formula expressing B in terms of Q and D.

5. Here is a trick that shows an interesting result when two odd numbers or two even numbers are paired.

Choose any two odd numbers or any two even numbers, such as 37 and 25.

Add the two numbers. $37 + 25 = 62$

Divide the sum by 2. $\dfrac{62}{2} = 31$

Subtract the two numbers. $37 - 25 = 12$

Divide the difference by 2. $\dfrac{12}{2} = 6$

Add the two quotients. $31 + 6 = 37$

The answer is always the larger of the two numbers.

Use algebra to explain this trick. There are two cases to consider. For the first case, that of two odd numbers, represent the numbers as $2K_1 + 1$ and $2K_2 + 1$, where K_1 and K_2 can represent any whole numbers. Assume K_1 is greater than K_2. For the second case, that of two even numbers, let $2K_1$ represent one number and $2K_2$ the other. Again assume K_1 is greater than K_2.

The Mysteries of
the Calendar

Finding Easter Sunday

We know that the calendar plays a very important part in the study of history, but we often overlook the mathematical aspects of this common device. Since a calendar designates days with numerals, it is dependent upon mathematics for its basic construction. Because of this, the calendar can be a source of many mathematical mysteries.

As you know, Easter Sunday has a way of being quite a movable date. Sometimes it occurs in March and sometimes in April. In 325 A.D., the Council of Nicaea established the rule that "Easter Day is always the first Sunday after the Full Moon, which happens upon or next after the twenty-first Day of March; and if the Full Moon happens on a Sunday, Easter Day is the Sunday after."

To determine the date for Easter Sunday in any year, the following items need to be considered:

1. The date of the first full moon following March 21.

2. The day of the week on which this full moon falls.

3. The date of the following Sunday.

Karl Friedrich Gauss (1777-1855), a great German mathematician, expressed the process mathematically in the following manner:

1. Determine constants m and n from this table.

CONSTANTS	YEARS			
	1582-1699	1700-1799	1800-1899	1900-2000
m	22	23	23	24
n	2	3	4	5

Thus for 1959, $m = 24$ and $n = 5$.

2. Let a be the remainder when the number of the year is divided by 4:

$$\frac{1959}{4} = 489 \ R3, \ a = 3$$

3. Let b be the remainder when it is divided by 7:

$$\frac{1959}{7} = 279 \ R6, \ b = 6$$

4. Let c be the remainder when it is divided by 19:

$$\frac{1959}{19} = 103 \ R2, \ c = 2$$

5. Let d be the remainder when $19c + m$ is divided by 30:

$$\frac{19(2) + 24}{30} = \frac{62}{30} = 2 \ R2, \ d = 2$$

6. Let e be the remainder when $2a + 4b + 6d + n$ is divided by 7:

$$\frac{2(3) + 4(6) + 6(2) + 5}{7} = \frac{6 + 24 + 12 + 5}{7} = \frac{47}{7} = 6 \ R5, \ e = 5$$

Then the Easter full moon comes d days after March 21, and Easter Day is the $(22 + d + e)$th day of March, or the $(d + e - 9)$th day of April. Thus for 1959, Easter Sunday was on the $(22 + 2 + 5)$-th day of March, or March 29. According to the formula, it could not have been in April, for $d + e - 9$ is a negative number.

What date will Easter Sunday be next year? Find the answer by using the mathematical steps given above and check this with a copy of next year's calendar. You can be glad that calendar makers give you the date of Easter so that you don't have to perform this complicated mathematical procedure. But it is interesting to know that there is a mathematical way to obtain the answer.

What Day of the Week Was It?

Do you know on what day of the week you were born? If you don't, it isn't too hard to find out. Just follow these directions:

Assign the months the following key numbers:

January	1	July	0
February	4	August	3
March	4	September	6
April	0	October	1
May	2	November	4
June	5	December	6

In leap years use 0 for the key number for January and 3 for the key number for February.

Suppose the date being considered is December 25, 1958. What day of the week was this?

1. Take the last two figures of the year. 58

2. Divide by 4 and disregard fractional remainders. $\frac{58}{4} = 14$

3. Take the key number for the month. December = 6

4. Take the day of the month. 25

5. Add the numbers in Steps 1, 2, 3, and 4. $58 + 14 + 6 + 25 = 103$

6. Divide the total found in Step 5 by 7. $\frac{103}{7} = 14$ and 5 remainder

The remainder gives the day of the week:

1 is Sunday, 2 is Monday, 3 is Tuesday, 4 is Wednesday, 5 is Thursday, 6 is Friday, 0 is Saturday.

Therefore, Christmas Day occurred on Thursday in 1958. You might still have an old calendar around to check the truth of this.

You can also use this method to find on what day certain famous events, such as the signing of the Declaration of Independence on July 4, 1776, took place.

However, for dates other than those in the 1900's, you need to add more to the total at Step 5.

For years in the 1700's add 4 more

1800's add 2 more
2000's add 6 more
2100's add 4 more

It should be said that this method works only for dates after September 15, 1752, the date the Gregorian Calendar, the one we still use, was adopted.

The proof of the dependability of this method is very complicated and requires a great deal of knowledge about the structure of the Gregorian Calendar. The parts of our calendar are put together in a rather irregular fashion which makes algebraic analysis difficult. If you are interested in building such a proof, however, first learn all you can about our calendar by referring to an encyclopedia or other reference book. Try to determine the significance of the key numbers for months and the corrections for different centuries. Then, restate the method proposed above, using letters for values that can vary. Good luck!

The Total of Dates in One Week

Not all the mathematics related to the calendar is as tedious as the process of finding the date of Easter Sunday. In fact, you can do some simple mathematical tricks with a calendar. For example, take a calendar and ask a friend to choose some full week of seven days. It can be any week in any month as long as there are a full seven days in the week.

Suppose the month chosen looks like the one shown below.

December						
S	M	T	W	TH	F	S
		1	2	3	4	5
6	7	8	9	10	11	12
13	14	15	16	17	18	19
20	21	22	23	24	25	26
27	28	29	30	31		

Figure 1

Have him tell you what number is the first day in the week he has chosen. Suppose it is 13. Then have him add the numbers of the days that appear on the calendar for that week.

While he is adding

$$13 + 14 + 15 + 16 + 17 + 18 + 19,$$

you can tell him the sum by adding 3 to the first date of this week,

$$13 + 3 = 16,$$

and multiplying this result by 7.

$$7 \times 16 = 112$$

He should find the sum to be 112.

Your ability to find the answer much faster than your friend should raise your status considerably. Of course, there is a simple mathematical explanation of why your method is correct.

Suppose the date given is D. Then the sum of the numbers of the days starting with D is

$$D + (D + 1)) + (D + 2) + (D + 3) + (D + 4) + (D + 5) + (D + 6),$$

which can be written as

$$7D + (1 + 2 + 3 + 4 + 5 + 6) = 7D + 21 = 7(D + 3).$$

Thus, we have proved that the sum can be found by adding 3 to the number of the first day and then multiplying that result by 7. This method can be used to find the sum of the numbers of any seven consecutive days that do not extend from one month to another.

Another Calendar Sum

Take a calendar and ask a friend to select any square arrangement of dates having three rows and three columns. Have him tell you the lowest number in the number square he selected, and while he is adding all nine dates, you can tell him the sum in very short order by adding 8 to the number given and multiplying this sum by 9.

December						
S	M	T	W	TH	F	S
		1	2	3	4	5
6	7	8	9	10	11	12
13	14	15	16	17	18	19
20	21	22	23	24	25	26
27	28	29	30	31		

Figure 2

If the square is the one circled in Figure 2, then the sum of

$$2 +3 +4 +9 +10 +11 +16 +17 +18$$

is

$$9(2 +8) =9(10) =90$$

It will undoubtedly take your friend much longer to tell you the sum is 90.

This is not too hard to prove. You can select any date, D, and then fill in the other dates in the square, like this:

D	$D +1$	$(D +2)$
$D +7$	$(D +1) +7$	$(D +2) +7$
$D +2(7)$	$(D +1) +2(7)$	$(D +2) +2(7)$

An easy way to obtain the total is to add each of the three columns as subtotals and then add these three subtotals:

Subtotal of Column 1:	$3D +3(7)$
Subtotal of Column 2:	$3D +3(7) +3$
Subtotal of Column 3: $+$	$3D +3(7) +3(2)$
Grand total:	$9D +9(7) +9$
This becomes:	$9(D +7 +1)$
$=$	$9(D +8)$

This describes the rule that you used to get the total in the example.

EXERCISE SET 4
Problems from the Calendar

1. On August 15, 1914, the Panama Canal was opened. On what day of the week did this event take place?

2. December 16, 1773, is the date of the Boston Tea Party. On what day of the week did the men of Boston dump British tea into the ocean?

3. Find the sum of the numbers of seven consecutive days on a calendar if the first day is December 10.

4. Find the sum of the dates in a 3-by-3 arrangement of numbers on a calendar if the smallest number in the arrangement is 8.

Time for
Clock Fun

The Impossible Isn't Too Hard

Since the calendar represents a measure of time, what is more natural than to think of using another measure of time — the clock or watch — for the next bit of mathematical fun?

One of the famous problems of antiquity that defied solution for centuries is that of trisecting an angle by the use of ruler and compasses alone. It has been proved that the operation is impossible to perform when the tools are to be the ruler and compasses alone, but many people are not aware that this restriction of instruments is the very thing that makes it impossible.

Of course, you are not going to solve an impossible problem, but instead, you will merely remove the restriction that makes it impossible. Instead of using compasses and a straight edge to trisect an angle, you will use your watch.

Set your watch at 12:00 and let one side of an angle, *A*, to be trisected be understood to lie on the watch hands in this first position. Rotate the minute hand clockwise until its new position

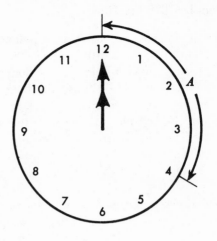

coincides with the second side of angle A. The angle through which the hour hand has rotated is $\frac{A}{12}$. We know this is true, for when the minute hand makes a complete rotation, the hour hand moves from one hour to the next, $\frac{1}{12}$ of a complete rotation. Multiply this angle, $\frac{A}{12}$, by 4 and you have $\frac{A}{3}$, or $\frac{1}{3}$ of angle A.

The distinctive feature in using a watch for this trick lies in the fact that the multiplication of $\frac{A}{12}$ by 4 can be done mechanically simply by rotating the minute hand on through three more angles the size of angle A. When the minute hand has rotated through $4A$, the hour hand has rotated $\frac{1}{12}$ as far, which is $\frac{1}{12}$ of $4A$ or $\frac{A}{3}$.

Experiment with this for several different angles.

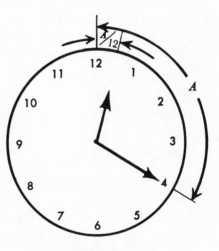

Figure 3

Tapping Out the Time

As long as you have your watch out, you might as well make use of it to perform a little magic — or at least what seems like magic.

Ask a friend to think of any hour on the face of a clock, such as 9:00. He doesn't tell you the hour he chooses. You start to tap on the face of the dial and ask your friend to count the taps to himself. Tell him to start with the hour he selected and add 1 at the first tap, 1 more at the second tap, and so on until he reaches 20. Thus, he will count to himself 10, 11, 12, . . ., 20 as you tap. Ask him to stop you when he reaches 20. At that instant, you will be pointing to the hour he chose, even though you had no way of knowing the starting number and therefore no way of knowing when you would be told to stop tapping.

To do this trick, you count the taps, too. You count the first 7 taps. On the eighth tap point to 12, and on each successive tap point to one number less: 11, 10, 9. When your friend says, "Stop!", you will be pointing to 9, the hour he picked at the beginning.

This seems very mysterious until you use mathematics to explain the mystery.

Your friend picked 9.

You count:	1	2	3	4	5	6	7	8				
You point to:								12	11	10	9	
Your friend counts:	10	11	12	13	14	15	16	17	18	19	20	Stop!

When he says, "Stop!" at 20 you have reached 9, the hour he picked.

As you examine the chart showing the counting, you will see that your friend actually counts (20-9) or eleven taps. Since the largest number he can select from the clock is 12, you know that you will be able to make at least eight taps before your friend counts 20. On your eighth tap, you put your finger on 12 on the clock. In the example, your friend had (11 −8) or three taps more to count. As you make each tap after the eighth one, you subtract 1 from 12 and arrive at 9 when he tells you to stop.

Put in general form, if the hour he selects is H, then he counts $(20 - H)$ taps. After he counts your first eight taps (which you also count), the number of taps that remain to be made is $(20 - H) - 8$. This represents the number which must be subtracted from 12 for the count-down to H.

In other words, $12 - [(20 - H) - 8]$ should give H, and it does by algebraic simplification:

$$12 - [(20 - H) - 8] =$$
$$12 - [20 - H - 8] =$$
$$12 - [12 - H] =$$
$$12 - 12 + H =$$
$$H$$

The Clock Tells More Than Time

Ask a friend to throw a die and to keep secret the number thrown. Have him also choose a number, say 10. The trick works faster if this number isn't too large, say under 50. He doesn't tell you what number he chooses.

Have him begin on the watch dial with the number shown by the die and tap clockwise ten times. The number on the watch dial reached on the tenth tap is written down.

Have him repeat the process tapping counter-clockwise. The number reached on the tenth tap is written down.

Have him now total these two numbers and give you the answer. You can immediately tell him what number he had on the die.

If the total is 12 or less, then the number on the die is half of this total. If the total is over 12, then you first subtract 12 and then take half this answer to give the number on the die.

Let's use the clock face in Figure 4 to illustrate.

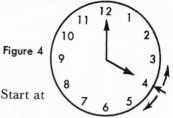

Figure 4

Suppose the number thrown was 4. Start at 4 on the clock face.

Count ten taps clockwise:

5	6	7	8	9	10	11	12	1	2
3	2	1	12	11	10	9	8	7	6

Count ten taps counter-clockwise:

Add 2 and 6: $2 + 6 = 8$

Divide by 2: $\dfrac{8}{2} = 4$

4 is the number on the die.

In the first part of the trick, your friend is really adding 4 and 10 on the clock face. You know that $4 + 10 = 14$, but the clock face numerals go only up to 12. When counting higher than 12 on a clock face, you must continue to count with 1, 2, and so on. The sum 14, expressed on the face of a clock, is the same as the remainder obtained after subtracting 12 from 14.

In the second part of the trick, your friend is using the clock face to subtract 10 from 4. But after subtracting 4 from 4 on the clock, he is pointing to the numeral 12. He then proceeds to subtract 6 (the remaining part of 10) from 12. This is equivalent to adding 12 to 4 and then subtracting 10.

In general, suppose the number obtained from the die is N and the number chosen is x. Then when you count x units clockwise, you arrive at the numeral $N + x - 12k$, where k represents the multiples of 12 subtracted because of the nature of the clock face.

Counting x units counter-clockwise, you arrive at the numeral $N + 12m - x$, where m represents the multiples of 12 added because of the nature of the clock face.

The sum of $N + x - 12k$ and $N + 12m - x$ is $2N - 12k + 12m$, or $2N + 12(m - k)$. You know that $N + x - 12k$ and $N + 12m - x$ are both equal to or less than 12, and hence their sum is equal to or less than 24. This indicates that $m - k$ must be either 0 or 1. If $m - k$ is 1, you will have to subtract 12 from the sum and then divide by 2 to get the number N. If $m - k$ is 0, dividing the sum by 2 will give the number N.

Mastering the Toys of Chance

Cards, Dice, and Dominoes

Since the preceding problem involved a die, you have already been introduced to one of the toys of chance. Gaming devices such as playing cards, dice, and even dominoes have long been props of the magician. Although many magicians may not know it, tricks with these instruments of chance can often be explained mathematically.

The Three Dice Sum

Bring out three dice and try this trick with a friend.

While your back is turned, have him throw the dice and ask him to add the values shown on the faces. Now have him pick up any one die and add the number on the bottom to the previous total. This same die is rolled again. Have your friend add the number it shows to the total.

Now turn around. Point out that you have no way of knowing what die he picked up and threw the second time, but, in spite of this, you can tell him the total he has obtained.

All you need to do is add 7 to what you see on the faces of the three dice.

This trick is easily explained by algebra.

Three dice are thrown. Let a, b, and c represent the values shown on the dice.

Add the numbers on the faces: $a + b + c$.

One die is picked up and the number of the bottom face is added. Note that opposite faces on a die always add to 7, so the number on the opposite face of a is $7 - a$, of b is $7 - b$, and of c is $7 - c$.

If a is the number shown on the die picked up, the new sum will be

$$(a+b+c)+(7-a)=b+c+7.$$

Since all three dice have the same characteristic number arrangement, the new sum will always be equal to 7 plus the values on the two dice not picked up.

Throwing the die again gives another face to add, say d, and the sum is

$$b+c+7+d.$$

What you see on the table is $b+c+d$.

What your friend has totaled is

$$b+c+d+7.$$

All you need to do then is add 7 to what you see to obtain your friend's total.

Mathematics and Playing Cards

Almost every amateur magician subjects his audience to card tricks. As a result, playing-card magic is often pretty dull. However, you should be able to hold the attention of your audience if you ask them to find the mathematical ideas that explain a trick.

Ask someone — let's call him Jim — to take from one to ten cards from a deck and hide them in his pocket without telling you the number. Then tell him to look at the card at that number from the top of the remainder of the deck and write the suit and denomination of the card on a slip of paper. Now ask another person to suggest a number greater than 10, but less than 40; suppose it is 18. Tell Jim to deal the cards one at a time on the table as he counts to 18. To demonstrate, take the cards and deal them from the top of the deck to form a pile face down on the table. Take one card for each count up to 18. Replace this pile of eighteen cards on the top of the deck.

Before Jim makes a second count of the eighteen cards, tell him to add to the top of the deck the cards he has in his pocket. Stress the fact that you have no way of knowing what that number of cards is. Nevertheless, after this second count of eighteen cards from the top of the deck, the card that will then be on the top of the remaining deck will be the card he originally selected.

Let's use mathematics to explain this trick. Let c represent the number of cards in the person's pocket and the position of the chosen card from the top of the deck. Let n be the number greater than 10 but less than 40 that was selected. Figure 5 represents this situation pictorially.

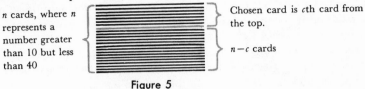

n cards, where n represents a number greater than 10 but less than 40

Chosen card is cth card from the top.

$n-c$ cards

Figure 5

When you demonstrate the counting of n cards from the top of the deck, you reverse the order of the n cards, as shown in Figure 6.

n cards

Chosen card is now $n-c$ cards from the top of the deck.

c cards

Figure 6

Adding the cards from the person's pocket places c cards on the top of the deck. This is demonstrated in Figure 7.

n cards

c cards

$n-c$ cards

c cards

Figure 7

We can see that the chosen card is $c+n-c$ or n cards below the top of the deck. After counting n cards the second time, the card remaining on the top of the deck is the chosen one.

Would the trick work if n was less than 10?

EXERCISE SET 5
The Three Stacked Dice Problem

Ask someone to take three dice and make a stack of them one on top of the other. You turn your back to him while he follows these instructions:

1. Look at the top and middle dice. Add the values shown on the two faces that touch each other.

2. Look at the bottom and middle dice. Add the values shown on the two faces that touch each other to the sum of Step 1.

3. Add the bottom face of the bottom die to the sum found in Step 2.

After your friend has found the answer, you turn around and tell him immediately what the final sum is. You can do this by subtracting the value of the top face of the top die from 21.

Why does this trick work?

Set up a proof in algebraic form, letting a and b represent the top and bottom faces respectively of the bottom die, c and d the top and bottom faces of the middle die, and e and f those of the top die.

A Domino Feat

Throwing dice is not the only way to pick a random number. As a matter of fact, a domino has the advantage of being more versatile than a die because it has two numbers on its face instead of just one.

Ask a friend to pick any domino. Suppose it is the one shown in Figure 8. Have him do the following arithmetic:

Multiply one of the numbers by 5:	$5(4) = 20$
Add 8:	$20 + 8 = 28$
Multiply by 2:	$2(28) = 56$
Add the other number on the domino:	$56 + 3 = 59$

Ask your friend for the answer.

Subtract 16 from his answer:	$59 - 16 = 43$
The two digits in the answer are the numbers on the domino:	4 and 3

Figure 8

Here is the proof by algebra:

Let the two numbers on the domino be x and y.

Multiply one number by 5:	$5x$
Add 8:	$5x + 8$
Multiply by 2:	$2(5x + 8) = 10x + 16$
Add the other number on the domino:	$(10x + 16) + y$
Subtract 16:	$10x + y$

This is a two-digit number because $10x$ places the x in the tens place, and y is in the ones place.

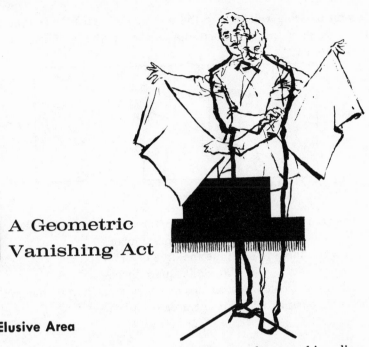

A Geometric Vanishing Act

An Elusive Area

Every good magician should be able to make something disappear and make something appear from nothing, and the mathematical magician is no exception. Some geometric figures are the only props needed for such mathematical mystery.

Draw a square sixteen inches by sixteen inches, and then draw lines to divide it into four parts, as shown in Figure 9.

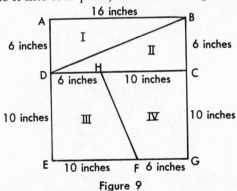

Figure 9

Now tell your audience that you will make a drawing that will rearrange the four parts of the square into a rectangle. Your second drawing will be like the one in Figure 10. Note that the dimensions of the four parts that make up the rectangle match perfectly.

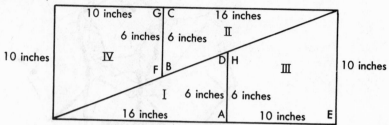

Figure 10

Ask your audience to compute the area of the square:

$$16 \times 16 = 256 \text{ square inches.}$$

Now ask them to compute the area of the rectangle:

$$10 \times 26 = 260 \text{ square inches.}$$

You have created four square inches of area. If you change the rectangle back to a square, the area disappears.

A Disappearing Line

Next, show your audience a circle similar to the one in Figure 11a. The circle is cut out so that it is possible to change its position. When the circle is in the position shown in Figure 11a with the arrow pointing to the numeral 1, twelve lines can be counted. When the circle is moved so that the arrow points to the numeral 2, as shown in Figure 11b, only eleven lines are visible. One line has disappeared.

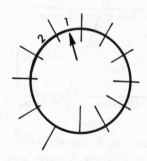

Figure 11a

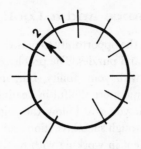

Figure 11 b

The branch of mathematics called geometry can be used to explain both tricks. Geometry, like every branch of mathematics, is a logical, well-organized field of study. We can establish most of the important ideas of geometry by starting with a few basic assumptions, ideas accepted as true without proof.

Assumptions play an important role in our disappearing tricks. In the first problem, four square inches of area seem to disappear because we make a false assumption. We assume that the lines *BD* and *FH* of the square will form a straight line when put together in the rectangle. They do not form a straight line. As a result, the parts do not fit together perfectly and a rectangle cannot be formed.

The disappearing line problem can be explained by a very basic geometric assumption. When the circle is moved from the first position to the second, the twelve line segments are rearranged to give only eleven line segments, each one a little longer than the original segments. The total length of the line segments in the first position is equal to the total length of the segments in the second position. The geometric principle applied is, "The whole is always equal to the sum of its parts, no matter how the parts are rearranged."

The two problems illustrate two types of difficulties encountered in the study of mathematics: sometimes we make assumptions that are false; sometimes we fail to recognize an application of a basic assumption.

A Backward Glance and a Look to the Future

Part VI has given you the opportunity to master a number of interesting mathematical tricks and puzzles. The problems given should enable you to impress people with your ability, but, more important, they should help you to become more skillful in dealing with mathematical situations. We hope that you will follow the suggestions to make up original problems, for through such invention and experimentation you will find greater enjoyment in working with mathematics.

Providing curious tricks is only one minor aspect of the important field of mathematics. Through mathematics, we can study and solve many types of important questions and problems. But every problem is, in a sense, a puzzle, and when you solve a problem, you are, in a sense, performing a trick. If you develop this attitude, all parts of the study of mathematics will be interesting, enjoyable, and entertaining.

Extending Your Knowledge

Now that your appetite for mathematical puzzles, games, and other recreations has been whetted, perhaps you would like to read other books with more material like this. Here is a brief list of such books.

ADLER, IRVING, *The Magic House of Numbers*. The John Day Co., 1957

BAKST, AARON, *Mathematical Puzzles and Pastimes*. D. Van Nostrand Co., 1954

DE GRAZIA, JOSEPH, *Math Is Fun*. Emerson Books, 1954

GARDNER, MARTIN, *Mathematics, Magic and Mystery*. Dover Publications, 1956

———————— *Mathematical Puzzles and Diversions*. Simon and Schuster, 1959

HEATH, ROYAL, *Mathemagic*. Dover Publications, 1953

HUNTER, J. A. H., *Fun with Figures*. Oxford University Press, 1958

KRAITCHIK, MAURICE, *Mathematical Recreations*. Dover Publications, 1942

MEYER, JEROME, *Fun with Mathematics*. Dover Publications, 1952

MOTT-SMITH, GEOFFREY, *Mathematical Puzzles for Beginners and Enthusiasts*. Dover Publications, 1954

SOLUTIONS TO THE EXERCISES

PART I

Solutions to the Exercises

EXERCISE SET 1

1. *a.* 3000   *b.* $\dfrac{3}{10,000}$ *c.* 3/10

2. 432

3. 234

4.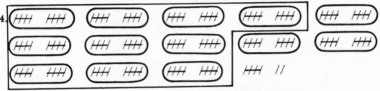

5. *a.* $(5 \times 10 \times 10 \times 10) + (6 \times 10 \times 10) + (8 \times 10) + (4 \times 1)$
   b. $(2 \times 10 \times 10 \times 10) + (0 \times 10 \times 10) + (7 \times 10) + (0 \times 1)$
   c. $\dfrac{4}{10} + \dfrac{1}{10 \times 10} + \dfrac{3}{10 \times 10 \times 10}$

6. *a.* Because we have 10 fingers.
   b. Because they used 10 fingers and 10 toes for counting.

EXERCISE SET 2

1. *a.*  *b.*

c. A group of ten groups of ten x's, three groups of ten x's, and seven x's

2. *a.*
b.
c.

3. *a.* $(5 \times 10) + 3$
   b. $(1 \times 10 \times 10) + (2 \times 10) + 3$
   c. $(2 \times 10 \times 10 \times 10) + (3 \times 10 \times 10) + (0 \times 10) + 4$

4. *a.* $(4 \times \text{five}) + 3$
   b. $(1 \times \text{five} \times \text{five}) + (2 \times \text{five}) + 3$
   c. $(2 \times \text{five} \times \text{five} \times \text{five}) + (3 \times \text{five} \times \text{five}) + (0 \times \text{five}) + 4$

5. *a.* $(2 \times \text{five}) + 4 = 14$
   b. $(4 \times \text{five} \times \text{five}) + (4 \times \text{five}) + 1 = 121$
   c. $(1 \times \text{five} \times \text{five} \times \text{five}) + (2 \times \text{five} \times \text{five}) + (0 \times \text{five}) + 3 = 178$

EXERCISE SET 3

1. *a.* 123_{five} *b.* 224_{five} *c.* 1111_{five} *d.* 3002_{five}

2. *a.* 13   *b.* 62   *c.* 95   *d.* 253

EXERCISE SET 4

1.
2	3	4	10	11
3	4	10	11	12
4	10	11	12	13
10	11	12	13	14
11	12	13	14	20

2. *a.* 44_five *b.* 41_five *c.* 112_five *d.* 1003_five

3. A ten   4. A five

EXERCISE SET 5

1. *a.* 21_five *b.* 13_five *c.* 223_five *d.* 3124_five

2. Ten   3. Five

EXERCISE SET 6

1. *a.* 20_five *b.* 30_five *c.* 31_five

2. *a.* 3   *b.* 3   *c.* 4

EXERCISE SET 7

1. *a.* 201_five *b.* 332_five *c.* 1442_five *d.* 4431_five

2. *a.* Yes   *b.* Yes   *c.* Yes

3. *a.* Yes   *b.* Yes   *c.* Yes

4. Yes

EXERCISE SET 8

1. 112_five 2. 40_five 3. 21_five 4. 104_five

EXERCISE SET 9

1. 53   2. 9   3. 43   4. 28

EXERCISE SET 10

1. *a.* 66_twelve *b.* $9T_\text{twelve}$ *c.* 311_twelve *d.* 2234_twelve

2. *a.* 92   *b.* 131   *c.* 377   *d.* 264

EXERCISE SET 11

1. *a.* 18_twelve *b.* 16_twelve *c.* $T6_\text{twelve}$ *d.* 100_twelve

2. *a.* $E5_\text{twelve}$ *b.* 407_twelve *c.* $9E38_\text{twelve}$ *d.* 1938_twelve

3. 15, 20, 27, 32, 39, 44, 4E

4. *a.* 9   *b.* 21_twelve *c.* 37_twelve *d.* $6E_\text{twelve}$

     e. 46_twelve *f.* $T4_\text{twelve}$ *g.* 123_twelve *h.* 17_twelve

EXERCISE SET 12

1. 16_twelve 2. $1T_\text{twelve}$ 3. 20_twelve

4. 5   5. 7   6. T

EXERCISE SET 13

1. *a.* 65_twelve *b.* $E3_\text{twelve}$ *c.* 186_twelve *d.* 653_twelve

     e. $29T_\text{twelve}$ *f.* 534_twelve *g.* 390_twelve *h.* 378_twelve

2. Yes

3. *a.* 38_twelve *b.* 537_twelve *c.* 65_twelve *d.* 262_twelve

EXERCISE SET 14

1. *a.* :8   *b.* 1:6   *c.* :T   *d.* :3   *e.* :4   *f.* :46

2. *a.* 5/12   *b.* 6/12 or   *c.* 8/12 or   *d.* 1/3   *e.* 28/144 or   *f.* 3 10/12 or

                     1/2   2/3   7/36   3⅚

3. 2, 3, 4, 6, 10_twelve

EXERCISE SET 15

1. *a.* 145_twelve inches   *b.* 27_twelve inches   *c.* 874_twelve inches   *d.* 4

2. $\dfrac{37_\text{twelve}}{100_\text{twelve}} = \dfrac{43}{144} = 29.9\%$

3. *a.* 12%   *b.* 8 per gross

EXERCISE SET 16

1. *a.* x

b.

c. x

d. x

2. *a.* 101   *b.* 1010   *c.* 1101   *d.* 1001

3. *a.* One twindred one
   b. One twosand twin
   c. One twosand one twindred one
   d. One twosand one

4. *a.*

b.  x

c.

d.

5. 10000, 10001, 10010, 10011, 10100, 10101, 10110, 10111, 11000, 11001, 11010, 11011, 11100, 11101, 11110, 11111, 100000, 100001, 100010, 100011

6. Small numbers use many digits.

EXERCISE SET 17

1. *a.* 21   *b.* 51   *c.* 89   *d.* 127
2. *a.* 10011_{two} *b.* 100101_{two} *c.* 1000001_{two} *d.* 1100100_{two}
3. 0

EXERCISE SET 18

1. *a.* 1001_{two} *b.* 1110_{two} *c.* 10100_{two} *d.* 100000_{two}
2. *a.* 101_{two} *b.* 10_{two} *c.* 10010_{two} *d.* 10_{two}
3. *a.* 110_{two} *b.* 1111_{two} *c.* 100011_{two} *d.* 111111_{two}
4. *a.* 1   *b.* 101_{two} *c.* 110_{two} *d.* $1001r101_{two}$

EXERCISE SET 19

1. Second   2. First

EXERCISE SET 20

1. *a.* 32, 8, 4, 2   *b.* 64, 32, 4, 2   *c.* 64, 32, 16
2. Cards like those in Figure 29: 5 cards, 16 numbers on each.
3. 1, 2, 4, 8

EXERCISE SET 21

1. If the units digit is 6 or 0, the number is divisible by 6.
2. Yes. If the sum of the last two digits is an even number, the number is divisible by 2.
3. If the sum of the digits is divisible by 4.
4. If the sum of the digits is divisible by 11(E).

EXERCISE SET 22

1. φ, ∀, △, φ□, φφ, φ∀, φ△, ∀□, ∀φ, ∀∀, ∀△, △□, △φ, △∀, △△, φ□□, φ□φ, φ□∀, φ□△, φφ□, φφφ, φφ∀, φφ△, φ∀□, φ∀φ.
2. *a.* φφ   *b.* ∀△   *c.* φ∀   *d.* ∀φφr△
3 and 4. Check your invention with a mathematics teacher.

NUMERATION REVIEW TEST

1. T	**2.** F	**3.** T	**4.** T
5. T	**10.** F	**11.** T	**12.** F
13. T	**14.** F	**15.** T	**20.** F
21. T	**22.** B	**23.** B	**24.** A
25. E	**30.** C	**31.** C	**32.** E
33. C	**34.** E	**35.** D	**40.** 10E
41. 11011_{two}	**42.** 3214_{five}	**43.** 2234_{six}	**44.** 328
45. 214	**50.** 43	**51.** 13014_{five}	**52.** 141_{five}
53. 30342_{five}	**54.** 14_{five}	**55.** $995E_{twelve}$	**100.** 9081_{twelve}
101. $3E314_{twelve}$	**102.** $85_{twelve}7$	**103.** 10001_{two}	**104.** 1110_{two}
105. 11110_{two}	**110.** 1101_{two}	**111.** 540_{six}	**112.** 444_{six}
113. 302_{six}	**114.** 24_{six}	**115.** :8	**120.** $\dfrac{1}{4}$
121. Six	**122.** Four	**123.** Six	

PART II

Solutions to the Exercises

EXERCISE SET 1

1. $10,000 + 1000 + 100 + 10 + 1 = 11111$
2. $100,000 - 1 = 99,999$
3. $(10)(10)(10)(10)(10) = 100,000$
4. $\dfrac{1}{(10)(10)(10)(10)(10)} = .00001$
5. $90,000 + 10,000 = 100,000$
6. $11(5) = 50 + 5$
7. $5 \times 9 = 50 - 5$
8. $5 \times 8 = 50 - 10$
9. $(12345 \times 9) + 6 = 111111$
10. $(12345 \times 8) + 5 = 98765$
11. $(54321 \times 9) - 1 = 488888$
12. $(54321 \times 8) - 1 = 434567$
13. $99999 \times 66666 = 6666533334$
14. $66667 \times 66667 = 4444488889$
15. $66666 \times 66667 = 4444422222$
16. $333333\ 666667 \times 111\ 3333 = 371111\ 371111\ 371111$
17. $333333\ 666667 \times 222\ 3333 = 741111\ 741111\ 741111$
18. $333333\ 666667 \times 111\ 6666 = 372222\ 372222\ 372222$
19. $333333\ 666667 \times 222\ 6666 = 742222\ 742222\ 742222$

EXERCISE SET 2

1. 121
2. 12321
3. 1234321
4. 123454321
5. 12345654321
6. 1234567654321
7. 123456787654321
8. 12345678987654321

EXERCISE SET 3

1. $11(5) = 10(5) + 5$ and, in general, $11(n) = 10(n) + n$
2. $9(5) = 10(5) - 5$ and, in general, $9(n) = 10(n) - n$
3. $8(5) = 10(5) - 2(5)$ and, in general, $8(n) = 10(n) - 2(n)$
4. $(5 + 1)^2 = 5^2 + 5 + 5 + 1$ and, in general, $(n + 1)^2 = n^2 + n + n + 1$
5. $5^2 + 5 = 6^2 - 6$ and, in general, $n^2 + n = (n + 1)^2 - (n + 1)$
6. $5(5) = \dfrac{5}{2}(10)$ and, in general, $n(5) = \dfrac{n}{2}(10)$
7. $5(25) = \dfrac{5}{4}(100)$ and, in general, $n(25) = \dfrac{n}{4}(100)$
8. $15(5) = 5(10) + \dfrac{5}{2}(10)$ and, in general, $15(n) = n(10) + \dfrac{n}{2}(10)$

EXERCISE SET 4

1. The right-hand column shows multiples of 143 × 7, or 11 × 13 × 7, which is 1001. The combination of the two columns gives, therefore,

$$
\begin{aligned}
(143 \times 1) \times 7 &= (11 \times 13 \times 7) \times 1 = 1001 \\
(143 \times 2) \times 7 &= (11 \times 13 \times 7) \times 2 = 2002 \\
(143 \times 3) \times 7 &= (11 \times 13 \times 7) \times 3 = 3003 \\
(143 \times 4) \times 7 &= (11 \times 13 \times 7) \times 4 = 4004 \\
(143 \times 5) \times 7 &= (11 \times 13 \times 7) \times 5 = 5005 \\
(143 \times 6) \times 7 &= (11 \times 13 \times 7) \times 6 = 6006 \\
(143 \times 7) \times 7 &= (11 \times 13 \times 7) \times 7 = 7007 \\
(143 \times 8) \times 7 &= (11 \times 13 \times 7) \times 8 = 8008 \\
(143 \times 9) \times 7 &= (11 \times 13 \times 7) \times 9 = 9009.
\end{aligned}
$$

2. All other fractions,

$$
\frac{444}{12}, \quad \frac{555}{15}, \quad \frac{666}{18}, \quad \frac{777}{21}, \quad \frac{888}{24}, \quad \frac{999}{27},
$$

are equal to 37. This is because each one can be changed to

$$
\frac{n(111)}{n(3)} = \frac{111}{3} = \frac{3 \times 37}{3} = 37.
$$

3. The combination of the two columns shows multiples of 15873 × 7, or 111,111. Therefore n times 111,111 shows n repeated in each digit.

$$
\begin{aligned}
(15873 \times 1) \times 7 &= (15873 \times 7) \times 1 = 111,111 \\
(15873 \times 2) \times 7 &= (15873 \times 7) \times 2 = 222,222 \\
(15873 \times 3) \times 7 &= (15873 \times 7) \times 3 = 333,333 \\
(15873 \times 4) \times 7 &= (15873 \times 7) \times 4 = 444,444 \\
(15873 \times 5) \times 7 &= (15873 \times 7) \times 5 = 555,555 \\
(15873 \times 6) \times 7 &= (15873 \times 7) \times 6 = 666,666 \\
(15873 \times 7) \times 7 &= (15873 \times 7) \times 7 = 777,777 \\
(15873 \times 8) \times 7 &= (15873 \times 7) \times 8 = 888,888 \\
(15873 \times 9) \times 7 &= (15873 \times 7) \times 9 = 999,999
\end{aligned}
$$

EXERCISE SET 5

1. The three expressions are identical for all n:

$$
1. \quad n\left(\frac{n}{n+1}\right) = \frac{n^2}{n+1}
$$

$$
2. \quad n - \frac{n}{n+1} = \frac{n^2 + n - n}{n+1} = \frac{n^2}{n+1}
$$

$$
3. \quad n - 1 + \frac{1}{n+1} = \frac{n^2 - 1 + 1}{n+1} = \frac{n^2}{n+1}
$$

2. The three expressions are identical for all n:

$$
1. \quad \left(n + \frac{1}{n+2}\right) \div \frac{n+1}{n+2} = \frac{n^2 + 2n + 1}{n+2} \times \frac{n+2}{n+1} = \frac{(n+1)^2}{n+1} = n+1
$$

$$
2. \quad \left(n + \frac{1}{n+2}\right) + \frac{n+1}{n+2} = \frac{n^2 + 2n + 1}{n+2} + \frac{n+1}{n+2} = \frac{n^2 + 3n + 2}{n+2}
$$

$$
= \frac{(n+1)(n+2)}{n+2} = n+1
$$

$$
3. \quad n + 1 = n + 1
$$

3. The three expressions are:

1. $\left(n + 2 + \dfrac{1}{n}\right) \div (n + 1) = \dfrac{n^2 + 2n + 1}{n} \times \dfrac{1}{n + 1} = \dfrac{(n + 1)^2}{n} \times \dfrac{1}{n + 1} = \dfrac{n + 1}{n}$

2. $\left(n + 2 + \dfrac{1}{n}\right) - (n + 1) = \dfrac{n^2 + 2n + 1}{n} - \dfrac{n^2 + n}{n} = \dfrac{n + 1}{n}$

3. $1 + \dfrac{1}{n} = \dfrac{n + 1}{n}$

EXERCISE SET 6

n	n^3	$n^3 - n$
2	8	$8 - 2 = 6$
3	27	$27 - 3 = 24$
4	64	$64 - 4 = 60$
5	125	$125 - 5 = 120$

6, 24, 60, 120 are all divisible by 6.

EXERCISE SET 7

1. $(4)(5)(6)(7)$ $= [(4)(7)] \; [(5)(6)]$
$= (28)(30)$
$= (29 - 1)(29 + 1)$
$= 29^2 - 1$

2. $(5)(6)(7)(8)$ $= [(5)(8)] \; [(6)(7)]$
$= (40)(42)$
$= (41 - 1)(41 + 1)$
$= 41^2 - 1$

3. $(9)(10)(11)(12) = [(9)(12)] \; [(10)(11)]$
$= (108)(110)$
$= (109 - 1)(109 + 1)$
$= 109^2 - 1$

EXERCISE SET 8

1. Start with any number: N
Multiply it by 3: $3N$
Subtract 1: $3N - 1$
Subtract 1: $3N - 2$
Add the three numbers: $3N + (3N - 1) + (3N - 2) = 9N - 3$
Adding the digits is equivalent to casting out 9's, so let's find the remainder when $9N - 3$ is divided by 9. We cannot have a negative remainder when dividing two positive numbers, so we write $9N - 3$ as $(9N - 9) + 6$. When this expression is divided by 9, the remainder is 6.

2. Take any two-digit number: $10a + b$
Reverse the digits: $10b + a$
Add the number and its reverse: $(10a + b) + (10b + a) = 11a + 11b$
This is divisible by 11 for all a's and b's.

EXERCISE SET 9

1. *a.* For $n = 1$, then $\dfrac{n(n + 1)}{2} = \dfrac{1(1 + 1)}{2} = \dfrac{1(2)}{2} = 1$,
which agrees with the first term, 1.

b. For $n = 2$, then $\dfrac{n(n + 1)}{2} = \dfrac{2(2 + 1)}{2} = \dfrac{2(3)}{2} = 3$,
which agrees with the sum of the first 2 terms: $1 + 2 = 3$.

c. For $n = 3$, then $\dfrac{n(n + 1)}{2} = \dfrac{3(3 + 1)}{2} = \dfrac{3(4)}{2} = 6,$

which agrees with the sum of the first 3 terms: $1 + 2 + 3 = 6.$

d. For $n = 4$, then $\dfrac{n(n + 1)}{2} = \dfrac{4(4 + 1)}{2} = \dfrac{4(5)}{2} = 10,$

which agrees with the sum of the first 4 terms: $1 + 2 + 3 + 4 = 10.$

2. *a.* For $n = 1$, then $n(n + 1) = 1(1 + 1) = 1(2) = 2,$
which agrees with the first term, 2.

   b. For $n = 2$, then $n(n + 1) = 2(2 + 1) = 2(3) = 6,$
which agrees with the sum of the first 2 terms: $2 + 4 = 6.$

   c. For $n + 3$, then $n(n + 1) = 3(3 + 1) = 3(4) = 12,$
which agrees with the sum of the first 3 terms: $2 + 4 + 6 = 12.$

   d. For $n = 4$, then $n(n + 1) = 4(4 + 1) = 4(5) = 20,$
which agrees with the sum of the first 4 terms: $2 + 4 + 6 + 8 = 20.$

3. *a.* For $n = 1$, then $n^2 = 1^2 = 1,$
which agrees with the first term, 1.

   b. For $n = 2$, then $n^2 = 2^2 = 4,$
which agrees with the sum: $1 + 2 + 1 = 4.$

   c. For $n = 3$, then $n^2 = 3^2 = 9,$
which agrees with the sum: $1 + 2 + 3 + 2 + 1 = 9.$

   d. For $n = 4$, then $n^2 = 4^2 = 16,$
which agrees with the sum: $1 + 2 + 3 + 4 + 3 + 2 + 1 = 16.$

(For problems 4 through 8, only one value is verified. You may check the others in a similar way.)

4. *b.* For $n = 2$, then $\dfrac{n(n + 1)(2n + 1)}{6} = \dfrac{2(2 + 1)(2 \times 2 + 1)}{6} = \dfrac{2(3)(5)}{6} = 5,$

which agrees with the sum of the first 2 terms: $1^2 + 2^2 = 1 + 4 = 5.$

5. *b.* For $n = 2$, then $\dfrac{n^2(n + 1)^2}{4} = \dfrac{2^2(2 + 1)^2}{4} = \dfrac{4(3)^2}{4} = 9,$

which agrees with the sum of the first 2 terms: $1^3 + 2^3 = 1 + 8 = 9.$

6. *b.* For $n = 2$, then $n^2(2n^2 - 1) = 2^2[2(2)^2 - 1]$
$$= 4(8 - 1)$$
$$= 4(7)$$
$$= 28,$$

which agrees with the sum of the first 2 terms: $1^3 + 3^3 = 1 + 27 = 28.$

7. *b.* For $n = 2$, then $\dfrac{n}{3}(n + 1)(n + 2) = \dfrac{2}{3}(2 + 1)(2 + 2)$
$$= \dfrac{2}{3}(3)(4)$$
$$= 8,$$

which agrees with the sum of the first 2 terms: $1(2) + 2(3) = 2 + 6 = 8.$

8. *b.* For $n = 2$, then $\dfrac{n}{n + 1} = \dfrac{2}{2 + 1} = \dfrac{2}{3},$

which agrees with the sum of the first 2 terms:

$$\dfrac{1}{1(2)} + \dfrac{1}{2(3)} = \dfrac{1}{2} + \dfrac{1}{6} = \dfrac{3}{6} + \dfrac{1}{6} = \dfrac{4}{6} = \dfrac{2}{3}.$$

EXERCISE SET 10

1. $n = 6$
   2% Simple: $1.10 + .02 = 1.12$
   2% Compound: $(1.1041)(1.02) = 1.1262$ (rounded off)
   4% Simple: $1.20 + .04 = 1.24$
   4% Compound: $(1.2167)(1.04) = 1.2654$ (rounded off)
   6% Simple: $1.30 + .06 = 1.36$
   6% Compound: $(1.3382)(1.06) = 1.4185$ (rounded off)

2. $n = 21$
   2% Simple: $1.40 + .02 = 1.42$
   2% Compound: $(1.4859)(1.02) = 1.5156$ (rounded off)
   4% Simple: $1.80 + .04 = 1.84$
   4% Compound: $(2.1911)(1.04) = 2.2787$ (rounded off)
   6% Simple: $2.20 + .06 = 2.26$
   6% Compound: $(3.2071)(1.06) = 3.3995$ (rounded off)

3. $n = 60$ (Use values in the table for $n = 50$ and $n = 10$.)
   2% Simple: $2.00 + .20 = 2.20$
   2% Compound: $(2.6916)(1.2190) = 3.2811$ (rounded off)
   4% Simple: $3.00 + .40 = 3.40$
   4% Compound: $(7.1067)(1.4802) = 10.5193$ (rounded off)
   6% Simple: $4.00 + .60 = 4.60$
   6% Compound: $(18.4202)(1.7908) = 32.9869$ (rounded off)

4. 8% interest

	Simple	Compound
$n = 1$	1.08	1.0800
$n = 2$	1.16	1.1664

5. 25 years
6. 18 years (to the nearest year)

EXERCISE SET 11

1.

6	1	8
7	5	3
2	9	4

4	9	2
3	5	7
8	1	6

2	9	4
7	5	3
6	1	8

6	7	2
1	5	9
8	3	4

4	3	8
9	5	1
2	7	6

8	3	4
1	5	9
6	7	2

2	7	6
9	5	1
4	3	8

2. $4(a + b + c + d) = 1 + 2 + \ldots + 16$
$$= \frac{1 + 16}{2} \times 16$$
$$= 17 \times 8$$
Therefore, $(a + b + c + d) = \dfrac{17 \times 8}{4} = 17 \times 2 = 34.$

PART III

Solutions to the Exercises

EXERCISE SET 1

The sets of numbers in Problems 5, 6, 8, 9, and 10 satisfy the Pythagorean Theorem.

EXERCISE SET 2

1. $1+3+5+ \ldots +23 = 144 = 12^2$
2. $(1+3+5+ \ldots +23)+25 = 144+25 = 169 = 13^2$
3. $144+25 = 169$ or $12^2+5^2 = 13^2$
4. $1+3+5+ \ldots +47 = 576 = 24^2$
   $(1+3+5+ \ldots +47)+49 = 576+49 = 625 = 25^2$
   $576+49 = 625$ or $24^2+7^2 = 25^2$
5. $1+3+5+ \ldots +79 = 1600 = 40^2$
   $(1+3+5+ \ldots +79)+81 = 1600+81 = 1681 = 41^2$
   $1600+81 = 1681$ or $40^2+9^2 = 41^2$

EXERCISE SET 3

1. $12^2 = 144$
2. $13^2 = 169$
3. $24^2 = 576$
4. $25^2 = 625$
5. $40^2 = 1,600$
6. $41^2 = 1,681$
7. $60^2 = 3,600$
8. $61^2 = 3,721$
9. $84^2 = 7,056$
10. $85^2 = 7,225$
11. $112^2 = 12,544$
12. $113^2 = 12,769$

EXERCISE SET 4

1. $m^2+\left\{\frac{1}{2}(m^2-1)\right\}^2 = m^2+\frac{1}{4}(m^4-2m^2+1) = m^2+\frac{1}{4}m^4-\frac{1}{2}m^2+\frac{1}{4} = \frac{1}{4}m^4+\frac{1}{2}m^2+\frac{1}{4}$

   $\left\{\frac{1}{2}(m^2+1)\right\}^2 = \frac{1}{4}(m^4+2m^2+1) = \frac{1}{4}m^4+\frac{1}{2}m^2+\frac{1}{4}$

   Thus $m^2+\left\{\frac{1}{2}(m^2-1)\right\}^2 = \left\{\frac{1}{2}(m^2+1)\right\}^2$, for they both can be changed to the same expression.

2. $m = 5 : 5, 12, 13$
   $m = 7 : 7, 24, 25$
   $m = 9 : 9, 40, 41$
   $m = 11 : 11, 60, 61$

4. $(ac-bd)^2+(ad+bc)^2 = a^2c^2-2acbd+b^2d^2+a^2d^2+2adbc+b^2c^2 =$
   $ a^2c^2+b^2d^2+a^2d^2+b^2c^2$

   $(a^2+b^2)(c^2+d^2) = a^2c^2+b^2d^2+a^2d^2+b^2c^2$

   Thus $(ac-bd)^2+(ad+bc)^2 = (a^2+b^2)(c^2+d^2)$, for they both can be changed to the same expression.

5. $(m^2-n^2)^2+(mn+mn)^2 = (m^2+n^2)(m^2+n^2)$
   or
   $(m^2-n^2)^2+(2mn)^2 = (m^2+n^2)^2$

6. (a) 3, 4, 5
   (b) 8, 6, 10
   (c) 5, 12, 13
   (d) 15, 8, 17
   (e) 12, 16, 20
   (f) 7, 24, 25

7. $(m^2-n^2)^2+(2mn)^2 = (m^2+n^2)^2$. Substitute $n=1$ and obtain:
   $(m^2-1)^2+(2m)^2 = (m^2+1)^2$. Divide by 4 and obtain:
   $\frac{(m^2-1)^2}{4}+m^2 = \frac{(m^2+1)^2}{4}$. Change this to the form:
   $\left\{\frac{1}{2}(m^2-1)\right\}^2+m^2 = \left\{\frac{1}{2}(m^2+1)\right\}^2$

EXERCISE SET 5

Problems **1-8** are all done exactly the same way as the example on Page 19. You just need to establish that in each case the three geometric figures are similar to one another.

9. Volume of water is the cross-sectional area of the pipe, πr^2, times the length, h.

Volume in the 3-inch pipe is: $\pi\left(\dfrac{3}{2}\right)^2 h = \left(\dfrac{\pi}{4}h\right)3^2$.

Volume in the 4-inch pipe is: $\pi\left(\dfrac{4}{2}\right)^2 h = \left(\dfrac{\pi}{4}h\right)4^2$.

Volume in the 5-inch pipe is: $\pi\left(\dfrac{5}{2}\right)^2 h = \left(\dfrac{\pi}{4}h\right)5^2$.

The combined volumes of the 3- and 4-inch pipes equal the volume of the 5-inch pipe because

$$\left(\dfrac{\pi}{4}h\right)3^2 + \left(\dfrac{\pi}{4}h\right)4^2 = \left(\dfrac{\pi}{4}h\right)5^2.$$

This reduces to $\qquad 3^2 + \qquad 4^2 = \qquad 5^2$ by dividing by $\left(\dfrac{\pi}{4}h\right)$.

EXERCISE SET 6

1. $\sqrt{5} = 2.2+$
2. $\sqrt{8} = 2.8+$
3. $\sqrt{10} = 3.2-$
4. $\sqrt{13} = 3.6+$
5. Let $h = $ the hypotenuse.
$$h^2 = (1)^2 + (\sqrt{N})^2$$
$$h^2 = 1 + \sqrt{N} \cdot \sqrt{N}$$
$$h^2 = 1 + N$$
$$h = \sqrt{1+N}$$
6. Let $h = $ the hypotenuse.
$$h^2 = (\sqrt{N})^2 + (\sqrt{M})^2$$
$$h^2 = \sqrt{N} \cdot \sqrt{N} + \sqrt{M} \cdot \sqrt{M}$$
$$h^2 = N + M$$
$$h = \sqrt{N+M}$$
7. The construction should show that $\sqrt{20} = 2\sqrt{5}$.

EXERCISE SET 7

1. $\sqrt{3} = 1.7+$
2. $\sqrt{8} = 2.8+$
3. $\sqrt{5} = 2.2+$
4. $\sqrt{7} = 2.6+$
5. $\sqrt{21} = 4.6-$

EXERCISE SET 8

1. 1.73
2. 2.24
3. 3.87
4. Yes. $(1.73)(2.24) = 3.8752$, or 3.88 to the nearest hundredth. This is within 1 hundredth and therefore checks.
5. Yes. $(1.73 + 1.41)(1.73 - 1.41) = (3.14)(.32) = 1.0048$ or 1.00 rounded off to the nearest hundredth.
6. $(\sqrt{5} + \sqrt{3})(\sqrt{5} - \sqrt{3}) = (\sqrt{5})^2 - (\sqrt{3})^2 = 5 - 3 = 2$
In general: $(\sqrt{N} + \sqrt{M})(\sqrt{N} - \sqrt{M}) = N - M$

EXERCISE SET 9

1. $c = 5.39$
2. $b = 4.58$
3. $a = 9.95$
4. $b = 6.93$
5. $c = 9.22$
6. $c = 7.07$
7. $b = 9.38$
8. The answers are doubled; tripled; 10 times as large; 100 times as large.

1. *a.* $AC = 15$ ft., $BC = 15\sqrt{3}$ ft. $= 25.98$ ft.
   b. AC and $BC = 15\sqrt{2}$ ft. $= 21.21$ ft.
2. *a.* $AB = 60$ ft., $BC = 30\sqrt{3}$ ft. $= 51.96$ ft.
   b. $AB = 30\sqrt{2}$ ft. $= 42.42$ ft., $BC = 30$ ft.
3. *a.* $AB = 20\sqrt{3}$ ft. $= 34.64$ ft., $AC = 10\sqrt{3}$ ft. $= 17.32$ ft.
   b. $AB = 30\sqrt{2}$ ft. $= 42.42$ ft., $AC = 30$ ft.
4. $AC = 16$ ft.
   $AB = 16\sqrt{2}$ ft. $= 22.62$ ft.
   $BD = 32$ ft.
   $CD = 16\sqrt{3}$ ft. $= 27.71$ ft.

PAGES 129-137

Measuring a Lake. $AB = 96'$
How Much Cable? $AC = 45'$
A Ladder Problem. $AC = 14.1'$
A Gate Problem. $AB = 6.7'$

Diamond Dimensions. Distance between home and second base is $90\sqrt{2}$ or 127.26 feet. Distance between pitcher's box and second base is $90\sqrt{2} - 60$ or 67.26 feet.
Distance between pitcher's box and first or third base is 63.7 feet.

Crossing the River. 5.4 miles per hour
Moving Machinery. $v = 4.5$ feet per second
A Problem in Forces. $F = 9.2$ pounds
A Rope Problem.
   $F_h = 40$ pounds
   $F_v = 40\sqrt{3}$ pounds $= 69.3$ pounds
Mathematics of a Wall Bracket.
   $BC = 30$
   $F_1 = 24$ pounds
   $F_2 = 40$ pounds
Scanning the Horizon: Aircraft Division. $x = 200$ miles
Scanning the Horizon: Down-to-Earth Division. $x = 3$ miles
The Stick and the Trunk. $AD = 42$ to the nearest inch
The Diameter of a Pipe. This is a 45° right triangle. Therefore, $\left(h + \dfrac{D}{2}\right)$ is equal to $\sqrt{2}\left(\dfrac{D}{2}\right)$. This simplifies to $2h + D = \sqrt{2}D$ or $D(\sqrt{2} - 1) = 2h$, or $D = \dfrac{2}{\sqrt{2} - 1}h$ and finally, $D = 4.828\,h$.

A Machinist's Brain Teaser. The Pythagorean Theorem applied to triangle OMB gives:
$$\left(\frac{D}{2} - h\right)^2 + \left(\frac{C}{2}\right)^2 = \left(\frac{D}{2}\right)^2$$
and this simplifies to:
$$\left(\frac{D^2}{4}\right) - Dh + h^2 + \frac{C^2}{4} = \frac{D^2}{4}$$
$Dh = h^2 + \dfrac{C^2}{4}$ and finally,
$$D = h + \frac{C^2}{4h}$$

A Problem with Gears. Since the figure is a 45° right triangle, then the following relationship holds:
$$(D - d) = \sqrt{2}d$$
and this simplifies to
$$D = d + \sqrt{2}d = (1 + \sqrt{2})d$$
$$= (1 + 1.414)d$$
$$= 2.414d$$

Milling a Bar. Triangle ABC is a 45° right triangle. $AB = \dfrac{D}{2}$, $x = AC$

Therefore, $x\sqrt{2} = \dfrac{D}{2}$, or $x = \left(\dfrac{\sqrt{2}}{4}\right)D$

But also, $x + h = \dfrac{D}{2}$, and eliminating x between these two equations gives
$$h = \frac{D}{2} - \frac{\sqrt{2}}{4}D$$
$$= \left(\frac{1}{2} - \frac{\sqrt{2}}{4}\right)D$$
$$= \frac{2 - \sqrt{2}}{4}D$$
$$= .146D$$

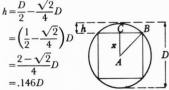

A Circle, a Hexagon, and a Triangle.
Triangle ADE is a 30°-60° right triangle. Therefore, $AD = \dfrac{\sqrt{3}}{2} r$ and $AC = \sqrt{3} \, r = 1.732 \, r$.

A Hexagon Fact. Triangle ABC is a 30°-60° right triangle with $AC = \dfrac{F}{2}$ and $BC = \dfrac{D}{4}$

Therefore, $\dfrac{F}{2} = \sqrt{3}\left(\dfrac{D}{4}\right)$

$$F = \dfrac{\sqrt{3}}{2} D$$
$$= \dfrac{1.732}{2} D = .866D$$

From a Square to an Octagon. Because of symmetry, it is sufficient to show that $EF = FG$.

Let $AB = x$. Then

$$AO = AF = \dfrac{\sqrt{2}}{2} x \text{ and } FB = x - \dfrac{\sqrt{2}}{2} x.$$

Since triangle FBG is a 45° right triangle, then

$$FG = \sqrt{2}\left(FB\right) = \sqrt{2}\left(x - \dfrac{\sqrt{2}}{2}x\right)$$
$$= \sqrt{2}x - x$$

It is also true that

$$EF = x - FB - AE = x - 2(FB)$$
$$= x - 2\left(x - \dfrac{\sqrt{2}}{2}x\right)$$
$$= x - 2x + \sqrt{2}x$$
$$= \sqrt{2}x - x$$

Thus EF and FG are both equal to the same expression in x and therefore equal to each other.

The Maltese Cross.

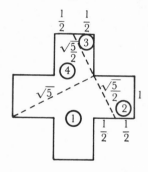

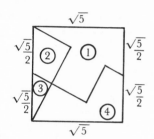

The Mystery of the Increased Area.
The pieces appear to make a triangle, but they really don't. An exaggerated drawing shows what is really happening.

The figure you put together is a triangle DCE of area $\dfrac{1}{2}(6)(8) = 24$ and a trapezoid of area

$$\dfrac{1}{2}(6 + 10)5 = \dfrac{1}{2}(16)5 = (8)5 = 40.$$

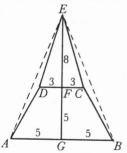

The total area is 64, the same as the original rectangle. The extra unit of area is in the two triangles ADE and BCE.

You can use the Pythagorean Theorem if you wish to compare AE with $AD + DE$. They are so nearly equal that you will have to carry out the square roots to several places before any difference is indicated.

PART IV

Solutions to the Exercises

EXERCISE SET 1

1. a. $\{a, e, i, o, u\}$; b. $\{1, 3, 5, 7, 9\}$; c. $\{$March, May$\}$
2. $\{2, 3, 5, 7, 11, 13, 17, 19\}$
3. Chevrolet, Buick, Pontiac
4. a. 26   b. 5
5. a. 9, 18, 27, 36, 45, 54, 63, 72, 81, 90
   b. Same as "a."
   c. Yes

EXERCISE SET 2

1. a. Maryland, Missouri, Minnesota, Montana, Maine, Massachusetts, Michigan, Mississippi
   b. $\{$I, V, X, L, C, D, M$\}$
   c. $\{$Jefferson, Jackson, Johnson$\}$
   d. $\{1, 4, 9, 16, 25, 36, 49, 64, 81\}$
2. a. the set of numbers less than 30 divisible by 5
   b. the set of numbers less than 20 divisible by 3
   c. the first 5 letters of our alphabet
   d. the days of the week beginning with T
3. a. Yes   b. No.   c. Yes   d. Yes

EXERCISE SET 3

1. Yes   2. Yes   3. No

EXERCISE SET 4

1. a. and d.   2. a.   3. a. 5;   b. 50;   c. 100;   d. 8
4. a., b., d.   5. a. 26;   b. 24;   c. 2;   d. 10

EXERCISE SET 5

1. d   2. a., b., d.   3. a., b., d.

EXERCISE SET 6

1. a. $\{3, 5, 7\}$ b. $\{3, 9\}$ c. $\{1, 3, 5, 7, 9\}$ d. $\{\ \}$
2. a. $\{x, y\}$ b. $\{$Ford$\}$ c. $\{$Christmas$\}$ d. $\{$squares$\}$ (Other examples are possible.)
3. a. $\{a, b\}, \{a, c\}, \{b, c\}, \{a\}, \{b\}, \{c\}, \{\ \}, \{a, b, c\}$
   b. $\{7, 11\}, \{7\}, \{11\}, \{\ \}$ c. List as for example 3a.
   d. 16 sets in all, $\{p, g\}, \{g, r\}$, etc.
4. 2, 4, 8, 16
5. a. 32   b. $P = 2^n$

EXERCISE SET 7

1. a. All students at Central High
   b. All American cars
   c. All types of sweaters
   d. Real numbers between 10 and 20
   (Other answers are possible.)
2. a. Any group of students in your school
   b. Numbers between 3 and 30
   c. Squares
   d. Mathematics books in your library
   (Other answers are possible.)

EXERCISE SET 8

1. a.  b.   c.

2. a.   b. ⬭ c. ◎ d. ○

3. a. IV   b. II   c. II   d. II

4. a.   b.   c.   d.

EXERCISE SET 9

1. a. $\{a, b, c, d, f, h\}$ b. $\{2, 3, 4, 6, 8, 9, 10, 12\}$
   c. $\{$Mary, Beth, Liz, Grace, Bonnie, Sue$\}$ d. $\{\triangle, \square, \bigcirc, \star\}$
2. a. 6. b. 8 c. 6 d. 4

3. a.   b. c. d.
4. a. A b. A c. U

EXERCISE SET 10

1. a. $\{a, e\}$ b. $\{$Plymouths$\}$ c. $\{\quad\}$ d. $\{3, 5, 7\}$

2. a.   b. c. d.
3. a. A b. ϕ c. A
4. $n(X \cup Y) = 14, 3$ 5. $n(X \cup Y) = n(X) + n(Y) - n(X \cap Y)$

EXERCISE SET 11

1. $\{6, 7, 8, 9\}$ 2. Not Cadillacs   3. All triangles not right triangles
4. a. 1960 Fords   b. Fords and all 1960 models   c. Not Fords
   d. 1960 Chevrolets, Plymouths, and Ramblers   e. Fords except 1960 Fords
   f. Chevrolets, Plymouths, Ramblers not 1960   g. Not 1960 Fords
   h. Same as "f"   i. Same as "g"

EXERCISE SET 12

1. a.   b.   c.   d. ◎

2. a.   b.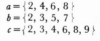

$a = \{2, 4, 6, 8\}$
$b = \{2, 3, 5, 7\}$
$c = \{2, 3, 4, 6, 8, 9\}$

c.

3. a. 20% b. 50% c. 10%
4. a. 1, 2, 3, 9 b. 3 c. 3
5. New York — Kansas City;
   Detroit — Baltimore;
   Chicago — Washington;
   Cleveland — Boston
6. a. 5/21 b. 16/21

EXERCISE 13

1.  s, u r, t A B v, x

2.  A B r, t

3. 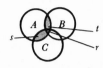 A B t s C r

4. 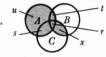 u t A B s r C x

EXERCISE SET 14

1. a.   b.   c.   d. $\{\ \}$

2. a. 3   b. 6   c. 3   d. 7

3. a.   b.   c.   d.

4. a. $\{A, B\}$ b.   c.

5. a. b. d.

EXERCISE SET 15

1. $\{5\}$ 2. $\{+5, -5\}$ 3. $\{6, 7, 8\}$ 4. $\{\ \}$ 5. $\{$all numbers$\}$

EXERCISE SET 16

1. a.   b.

   c.   d. All numbers possible.

2. a.

   b.

   c.

   d.

EXERCISE SET 17

1. a. $\{(2, 3), (1, 4), (0, 5)\}$ b. $\{(1, 10), (3, 6) (6, 0)\}$
   c. $\{(3, 2), (5, 8), (0, -7)\}$
2. a. $(4, -2)$ b. $(6, 4)$

EXERCISE SET 18

1. a. subtract 5   b. divide by 3   c. multiply by 1 or add 0
   d. add 1   e. none
2. a, b, c, d

EXERCISE SET 19

1. a.

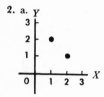

b.

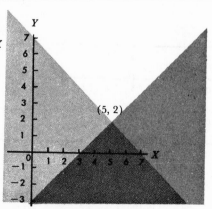

2. a.

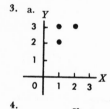

b.

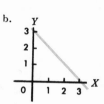

3. a.

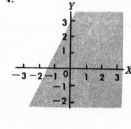

b.

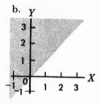

4.

5. $y = 2$ gals.

294

EXERCISE SET 20

1. a.   b. $\left(E\right)$ c.

2. a.

   b.

   a = All people interested
       in mathematics.
   b = All boys.
   c = All "A" mathematics students.

   a = Set of students who own cars.
   b = Set of students who get "A" grades.
   c = Set of girls.

   c.

   d.

   a = Set of all boys.
   b = Set of all St. Mark's students.
   c = Set of all courteous students.

   a = Set of all girls.
   b = Set of politicians.
   c = Set of all honest people.

3. a.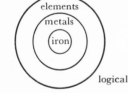

   b.

4. "Some freshmen are dumb bunnies" is not logical.

EXERCISE SET 21

1. F   2. T   3. F   4. F   5. T   6. T   7. T

8. T   9. T   10. T   11. ⬤⬤   12. ⬤

13. ◯ ◯   14. ▭⬤   15. ⬤⬤   16. ⬤⬤⬤

17. (A)(B)   18. (C)(D)   19. (X)(Z)(Y)   20. {2, 3, 4}

21. {2, 4}   22. {1, 3, 5, 7, 9}   23. {6, 8}

24. {1, 2, ... 6, 8}   25. ϕ

Solutions to the Exercises

EXERCISE SET 1

1. *a, c*   **2.** *a, b*   **3.** *a*—2; *b*—3; *c*—0; *d*—5

EXERCISE SET 2

Number of half-twists	Number of sides and edges	Results of cut
0	2	2 separate loops
1	1	1 loop, 2 twists
1	1	2 loops interlocked
2	2	2 loops interlocked
2	2	2 loops interlocked
3	1	1 loop, 1 knot
3	1	2 loops interlocked, 1 knot

EXERCISE SET 3

Figure	Even Vertices	Odd Vertices	Traveled
1.	2	2	Yes
2.	0	6	No
3.	4	0	Yes
4.	1	4	No
5.	10	0	Yes
6.	9	4	No
7.	4	8	No
8.	10	0	Yes
9.	2	2	Yes
10.	8	0	Yes
11.	2	6	No
12.	14	2	Yes

EXERCISE SET 4

1. Impossible   **2.** Impossible

EXERCISE SET 5

	V	*A*	*R*
1.	2	1	1
2.	2	2	2
3.	2	3	3
4.	4	6	4
5.	5	8	5
6.	4	7	5

EXERCISE SET 6

1. *b.* 11   *c.* 4   *d.* Kentucky, Pennsylvania, Colorado
2. Many arrangements of color are possible.
4. No. of regions: 2, 4, 7, 11, 16
    Differences: 2, 3, 4, 5
    With six lines: 22 regions

EXERCISE SET 7

1. *a*—2; *b*—3; *c*—3; *d*—0
2. *a.* singly; *b.* singly; *c.* doubly; *d.* triply; *e.* doubly; *f.* simple; *g.* simple

EXERCISE SET 8

	F	*E*	*V*	*V+F*	*E+2*
1.	4	6	4	8	8
2.	6	12	8	14	14
3.	8	12	6	14	14
4.	12	30	20	32	32
5.	20	30	12	32	32

1. *a.*
   b.
   c.
   d.
   e.

2. *a, b, c*
3. *a, b, c*
4. One
5. Any odd vertex
6. Two
7.

V	A	R
4	6	4
8	12	6
9	16	9
12	18	8
12	12	2

PUZZLE SOLUTIONS

Page 227. De-vesting Yourself

You can remove the vest by putting your left hand and arm through the left vest armhole. Then bring the entire coat and the right arm through this vest armhole. Finally, stuff the vest down the right coat sleeve and it will be off.

Page 227. Stringing Along

Loop your string under the wrist loop of your partner. Pull your loop over his hand and you will be free.

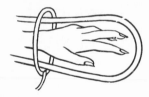

Page 228. Buttons and Beads

As in "Stringing Along," loop the center strings through the end holes and over the buttons.

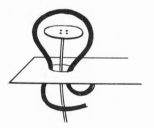

Page 228. The Swiss School Problem

Go around the houses.

Page 229. The Divided Farm

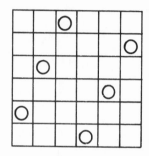

Page 229. Pennies on the Square

Page 229. The Street Sweeper's Route

Begin at any odd vertex. You will need to retrace three blocks.

Page 229. Paper and String Puzzle

Curve the paper as shown below and push the strip between the two slits through the hole from back to front. Either one of the buttons can then be passed through the loop formed by the center strip, and the string can easily be removed.

Page 230. The Paper Boot Puzzle

The boots can be removed by reversing the steps described. First refold the large piece of paper. Then slide the small piece back around point *A* and into position as shown in Figure 41*d*. The boots can then be lifted from the arm over which they are suspended.

PART VI

Solutions to the Exercises

EXERCISE SET 1

1. Choose a number: N

   Add 11: $N + 11$

   Multiply by 6: $6(N + 11) = 6N + 66$

   Subtract 3: $6N + 66 - 3 = 6N + 63$

   Divide by 3: $\dfrac{6N + 63}{3} = 2N + 21$

   Subtract 6 less than the
   original number: $2N + 21 - (N - 6) = 2N + 21 - N + 6$
   $$= N + 27$$

   Subtract 1 more than the
   original number: $N + 27 - (N + 1) = N + 27 - N - 1$
   $$= 26$$

   Divide by 2: $\dfrac{26}{2} = 13$

   The answer is always 13.

2. Take any number: N

   Multiply it by 2: $2N$

   Add 16 to that result: $2N + 16$

   Divide by 2: $\dfrac{2N + 16}{2} = N + 8$

   Subtract the original number: $N + 8 - N = 8$

   The answer is always 8.

EXERCISE SET 2

1. Multiply your age by 2: $2A$

   Add 10: $2A + 10$

   Multiply by 5: $5(2A + 10) = 10A + 50$

   Add the number of people
   in your family: $10A + 50 + N$

   Subtract 50: $10A + 50 + N - 50 = 10A + N$

   $10A$ leaves the units place for the number in the family. If there are more than 9 in the family, the units place won't be enough.

2. Double your house number: $2N$

   Add the number of days in
   a week: $2N + 7$

   Multiply by 50: $50(2N + 7) = 100N + 350$

   Add your age: $100N + 350 + A$

   Subtract 365: $100N + 350 + A - 365 = 100N + A - 15$

   Add 15: $100N + A - 15 + 15 = 100N + A$

   $100N$ puts the house number in the hundreds and higher places, leaving the tens and units places for the age.

EXERCISE SET 3

1. Take any number: N

   Add 4: $N + 4$

   Multiply by 3: $3(N + 4) = 3N + 12$

   Subtract 9: $3N + 12 - 9 = 3N + 3$

   Multiply by 3: $3(3N + 3) = 9N + 9$

   Add the original number: $9N + 9 + N = 10N + 9$

   9 is the units digit. Scratching it out leaves N.

2. $N + (N + 1) + (N + 2) + (N + 3) + (N + 4) + (N + 5) + (N + 6) = T$

   $7N + 21 = T$

   $7N = T - 21$

   $N = \dfrac{T - 21}{7}$

   Translated, this says that the number N is found by subtracting 21 from the total of the seven numbers and then dividing this difference by 7.

3. $N + (N + 1) + (N + 2) + \cdots + (N + x) = T$

   $(x + 1)N + (1 + 2 + \cdots + x) = T$

   The sum of the first x consecutive integers is given by the formula $\dfrac{x(x + 1)}{2}$.

   Therefore, the relation becomes:

   $(x + 1)N + \dfrac{x(x + 1)}{2} = T$

   $(x + 1)N = T - \dfrac{x(x + 1)}{2}$

   $N = \dfrac{T - \dfrac{x(x + 1)}{2}}{x + 1}$

4. $\dfrac{B}{A} = Q \qquad (1)$

   $B - A = D \qquad (2)$

   From (1), $A = \dfrac{B}{Q}$.

   Substitute $\dfrac{B}{Q}$ for A in (2)

   $B - \dfrac{B}{Q} = D$

   $B\left(1 - \dfrac{1}{Q}\right) = D$

   $B\left(\dfrac{Q - 1}{Q}\right) = D$

   $B = \dfrac{DQ}{Q - 1}$

5. *Case 1.*

Two odd numbers:	$2K_1 + 1$ and $2K_2 + 1$, where $K_1 > K_2$
Add the two numbers:	$(2K_1 + 1) + (2K_2 + 1)$ $= 2K_1 + 2K_2 + 2$
Divide the sum by 2:	$\dfrac{2K_1 + 2K_2 + 2}{2} = K_1 + K_2 + 1$
Subtract the two numbers:	$(2K_1 + 1) - (2K_2 + 1)$ $= 2K_1 - 2K_2$
Divide the difference by 2:	$\dfrac{2K_1 - 2K_2}{2}$ $= K_1 - K_2$
Add the two quotients:	$(K_1 + K_2 + 1) + (K_1 - K_2)$ $= 2K_1 + 1$

   This is the larger of the two original numbers.

Case 2.

Two even numbers:	$2K_1$ and $2K_2$,
	where $K_1 > K_2$
Add the two numbers:	$2K_1 + 2K_2$
Divide the sum by 2:	$\dfrac{2K_1 + 2K_2}{2} = K_1 + K_2$
Subtract the two numbers:	$2K_1 - 2K_2$
Divide the difference by 2:	$\dfrac{2K_1 - 2K_2}{2} = K_1 - K_2$
Add the two quotients:	$(K_1 + K_2) + (K_1 - K_2) = 2K_1$

This is the larger of the two original numbers.

EXERCISE SET 4

1. Saturday
2. Thursday
3. $7(10 + 3) = 91$
4. $9(8 + 8) = 144$

EXERCISE SET 5

Opposite faces of a die add up to 7. Therefore, $a + b = 7, c + d = 7$, and $e + f = 7$. The sum of all faces is 21.
$$(a + b) + (c + d) + (e + f) = 21,$$
and the sum of the faces your friend has added is $(f+c) + (d+a) + b$. But from the formula above, this is $21 - e$, where e is the value of the top face of the top die.

TABLE OF SQUARES AND SQUARE ROOTS

N	N²	√N	N	N²	√N	N	N²	√N
1	1	1.000	51	26 01	7.141	101	1 02 01	10.050
2	4	1.414	52	27 04	7.211	102	1 04 04	10.100
3	9	1.732	53	28 09	7.280	103	1 06 09	10.149
4	16	2.000	54	29 16	7.348	104	1 08 16	10.198
5	25	2.236	55	30 25	7.416	105	1 10 25	10.247
6	36	2.449	56	31 36	7.483	106	1 12 36	10.296
7	49	2.646	57	32 49	7.550	107	1 14 49	10.344
8	64	2.828	58	33 64	7.616	108	1 16 64	10.392
9	81	3.000	59	34 81	7.681	109	1 18 81	10.440
10	1 00	3.162	60	36 00	7.746	110	1 21 00	10.488
11	1 21	3.317	61	37 21	7.810	111	1 23 21	10.536
12	1 44	3.464	62	38 44	7.874	112	1 25 44	10.583
13	1 69	3.606	63	39 69	7.937	113	1 27 69	10.630
14	1 96	3.742	64	40 96	8.000	114	1 29 96	10.677
15	2 25	3.873	65	42 25	8.062	115	1 32 25	10.724
16	2 56	4.000	66	43 56	8.124	116	1 34 56	10.770
17	2 89	4.123	67	44 89	8.185	117	1 36 89	10.817
18	3 24	4.243	68	46 24	8.246	118	1 39 24	10.863
19	3 61	4.359	69	47 61	8.307	119	1 41 61	10.909
20	4 00	4.472	70	49 00	8.367	120	1 44 00	10.954
21	4 41	4.583	71	50 41	8.426	121	1 46 41	11.000
22	4 84	4.690	72	51 84	8.485	122	1 48 84	11.045
23	5 29	4.796	73	53 29	8.544	123	1 51 29	11.091
24	5 76	4.899	74	54 76	8.602	124	1 53 76	11.136
25	6 25	5.000	75	56 25	8.660	125	1 56 25	11.180
26	6 76	5.099	76	57 76	8.718	126	1 58 76	11.225
27	7 29	5.196	77	59 29	8.775	127	1 61 29	11.269
28	7 84	5.292	78	60 84	8.832	128	1 63 84	11.314
29	8 41	5.385	79	62 41	8.888	129	1 66 41	11.358
30	9 00	5.477	80	64 00	8.944	130	1 69 00	11.402
31	9 61	5.568	81	65 61	9.000	131	1 71 61	11.446
32	10 24	5.657	82	67 24	9.055	132	1 74 24	11.489
33	10 89	5.745	83	68 89	9.110	133	1 76 89	11.533
34	11 56	5.831	84	70 56	9.165	134	1 79 56	11.576
35	12 25	5.916	85	72 25	9.220	135	1 82 25	11.619
36	12 96	6.000	86	73 96	9.274	136	1 84 96	11.662
37	13 69	6.083	87	75 69	9.327	137	1 87 69	11.705
38	14 44	6.164	88	77 44	9.381	138	1 90 44	11.747
39	15 21	6.245	89	79 21	9.434	139	1 93 21	11.790
40	16 00	6.325	90	81 00	9.487	140	1 96 00	11.832
41	16 81	6.403	91	82 81	9.539	141	1 98 81	11.874
42	17 64	6.481	92	84 64	9.592	142	2 01 64	11.916
43	18 49	6.557	93	86 49	9.644	143	2 04 49	11.958
44	19 36	6.633	94	88 36	9.695	144	2 07 36	12.000
45	20 25	6.708	95	90 25	9.747	145	2 10 25	12.042
46	21 16	6.782	96	92 16	9.798	146	2 13 16	12.083
47	22 09	6.856	97	94 09	9.849	147	2 16 09	12.124
48	23 04	6.928	98	96 04	9.899	148	2 19 04	12.166
49	24 01	7.000	99	98 01	9.950	149	2 22 01	12.207
50	25 00	7.071	100	1 00 00	10.000	150	2 25 00	12.247

A CATALOGUE OF SELECTED DOVER BOOKS
IN ALL FIELDS OF INTEREST

A CATALOGUE OF SELECTED DOVER BOOKS
IN ALL FIELDS OF INTEREST

AMERICA'S OLD MASTERS, James T. Flexner. Four men emerged unexpectedly from provincial 18th century America to leadership in European art: Benjamin West, J. S. Copley, C. R. Peale, Gilbert Stuart. Brilliant coverage of lives and contributions. Revised, 1967 edition. 69 plates. 365pp. of text.
21806-6 Paperbound $3.00

FIRST FLOWERS OF OUR WILDERNESS: AMERICAN PAINTING, THE COLONIAL PERIOD, James T. Flexner. Painters, and regional painting traditions from earliest Colonial times up to the emergence of Copley, West and Peale Sr., Foster, Gustavus Hesselius, Feke, John Smibert and many anonymous painters in the primitive manner. Engaging presentation, with 162 illustrations. xxii + 368pp.
22180-6 Paperbound $3.50

THE LIGHT OF DISTANT SKIES: AMERICAN PAINTING, 1760-1835, James T. Flexner. The great generation of early American painters goes to Europe to learn and to teach: West, Copley, Gilbert Stuart and others. Allston, Trumbull, Morse; also contemporary American painters—primitives, derivatives, academics—who remained in America. 102 illustrations. xiii + 306pp. 22179-2 Paperbound $3.00

A HISTORY OF THE RISE AND PROGRESS OF THE ARTS OF DESIGN IN THE UNITED STATES, William Dunlap. Much the richest mine of information on early American painters, sculptors, architects, engravers, miniaturists, etc. The only source of information for scores of artists, the major primary source for many others. Unabridged reprint of rare original 1834 edition, with new introduction by James T. Flexner, and 394 new illustrations. Edited by Rita Weiss. 6⅝ x 9⅝.
21695-0, 21696-9, 21697-7 Three volumes, Paperbound $13.50

EPOCHS OF CHINESE AND JAPANESE ART, Ernest F. Fenollosa. From primitive Chinese art to the 20th century, thorough history, explanation of every important art period and form, including Japanese woodcuts; main stress on China and Japan, but Tibet, Korea also included. Still unexcelled for its detailed, rich coverage of cultural background, aesthetic elements, diffusion studies, particularly of the historical period. 2nd, 1913 edition. 242 illustrations. lii + 439pp. of text.
20364-6, 20365-4 Two volumes, Paperbound $6.00

THE GENTLE ART OF MAKING ENEMIES, James A. M. Whistler. Greatest wit of his day deflates Oscar Wilde, Ruskin, Swinburne; strikes back at inane critics, exhibitions, art journalism; aesthetics of impressionist revolution in most striking form. Highly readable classic by great painter. Reproduction of edition designed by Whistler. Introduction by Alfred Werner. xxxvi + 334pp.
21875-9 Paperbound $2.50

Two Little Savages; Being the Adventures of Two Boys Who Lived as Indians and What They Learned, Ernest Thompson Seton. Great classic of nature and boyhood provides a vast range of woodlore in most palatable form, a genuinely entertaining story. Two farm boys build a teepee in woods and live in it for a month, working out Indian solutions to living problems, star lore, birds and animals, plants, etc. 293 illustrations. vii + 286pp.

20985-7 Paperbound $2.50

Peter Piper's Practical Principles of Plain & Perfect Pronunciation. Alliterative jingles and tongue-twisters of surprising charm, that made their first appearance in America about 1830. Republished in full with the spirited woodcut illustrations from this earliest American edition. 32pp. $4\frac{1}{2}$ x $6\frac{3}{8}$.

22560-7 Paperbound $1.00

Science Experiments and Amusements for Children, Charles Vivian. 73 easy experiments, requiring only materials found at home or easily available, such as candles, coins, steel wool, etc.; illustrate basic phenomena like vacuum, simple chemical reaction, etc. All safe. Modern, well-planned. Formerly *Science Games for Children*. 102 photos, numerous drawings. 96pp. $6\frac{1}{8}$ x $9\frac{1}{4}$.

21856-2 Paperbound $1.25

An Introduction to Chess Moves and Tactics Simply Explained, Leonard Barden. Informal intermediate introduction, quite strong in explaining reasons for moves. Covers basic material, tactics, important openings, traps, positional play in middle game, end game. Attempts to isolate patterns and recurrent configurations. Formerly *Chess*. 58 figures. 102pp. (USO) 21210-6 Paperbound $1.25

Lasker's Manual of Chess, Dr. Emanuel Lasker. Lasker was not only one of the five great World Champions, he was also one of the ablest expositors, theorists, and analysts. In many ways, his Manual, permeated with his philosophy of battle, filled with keen insights, is one of the greatest works ever written on chess. Filled with analyzed games by the great players. A single-volume library that will profit almost any chess player, beginner or master. 308 diagrams. xli x 349pp.

20640-8 Paperbound $2.75

The Master Book of Mathematical Recreations, Fred Schuh. In opinion of many the finest work ever prepared on mathematical puzzles, stunts, recreations; exhaustively thorough explanations of mathematics involved, analysis of effects, citation of puzzles and games. Mathematics involved is elementary. Translated by F. Göbel. 194 figures. xxiv + 430pp. 22134-2 Paperbound $3.00

Mathematics, Magic and Mystery, Martin Gardner. Puzzle editor for Scientific American explains mathematics behind various mystifying tricks: card tricks, stage "mind reading," coin and match tricks, counting out games, geometric dissections, etc. Probability sets, theory of numbers clearly explained. Also provides more than 400 tricks, guaranteed to work, that you can do. 135 illustrations. xii + 176pp.

20338-2 Paperbound $1.50

PLANETS, STARS AND GALAXIES: DESCRIPTIVE ASTRONOMY FOR BEGINNERS, A. E. Fanning. Comprehensive introductory survey of astronomy: the sun, solar system, stars, galaxies, universe, cosmology; up-to-date, including quasars, radio stars, etc. Preface by Prof. Donald Menzel. 24pp. of photographs. 189pp. 5¼ x 8¼.
21680-2 Paperbound $1.50

TEACH YOURSELF CALCULUS, P. Abbott. With a good background in algebra and trig, you can teach yourself calculus with this book. Simple, straightforward introduction to functions of all kinds, integration, differentiation, series, etc. "Students who are beginning to study calculus method will derive great help from this book." Faraday House Journal. 308pp.
20683-1 Clothbound $2.00

TEACH YOURSELF TRIGONOMETRY, P. Abbott. Geometrical foundations, indices and logarithms, ratios, angles, circular measure, etc. are presented in this sound, easy-to-use text. Excellent for the beginner or as a brush up, this text carries the student through the solution of triangles. 204pp.
20682-3 Clothbound $2.00

TEACH YOURSELF ANATOMY, David LeVay. Accurate, inclusive, profusely illustrated account of structure, skeleton, abdomen, muscles, nervous system, glands, brain, reproductive organs, evolution. "Quite the best and most readable account,' *Medical Officer.* 12 color plates. 164 figures. 311pp. 4¾ x 7.
21651-9 Clothbound $2.50

TEACH YOURSELF PHYSIOLOGY, David LeVay. Anatomical, biochemical bases; digestive, nervous, endocrine systems; metabolism; respiration; muscle; excretion; temperature control; reproduction. "Good elementary exposition," *The Lancet.* 6 color plates. 44 illustrations. 208pp. 4¼ x 7.
21658-6 Clothbound $2.50

THE FRIENDLY STARS, Martha Evans Martin. Classic has taught naked-eye observation of stars, planets to hundreds of thousands, still not surpassed for charm, lucidity, adequacy. Completely updated by Professor Donald H. Menzel, Harvard Observatory. 25 illustrations. 16 x 30 chart. x + 147pp.
21099-5 Paperbound $1.25

MUSIC OF THE SPHERES: THE MATERIAL UNIVERSE FROM ATOM TO QUASAR, SIMPLY EXPLAINED, Guy Murchie. Extremely broad, brilliantly written popular account begins with the solar system and reaches to dividing line between matter and nonmatter; latest understandings presented with exceptional clarity. Volume One: Planets, stars, galaxies, cosmology, geology, celestial mechanics, latest astronomical discoveries; Volume Two: Matter, atoms, waves, radiation, relativity, chemical action, heat, nuclear energy, quantum theory, music, light, color, probability, antimatter, antigravity, and similar topics. 319 figures. 1967 (second) edition. Total of xx + 644pp.
21809-0, 21810-4 Two volumes, Paperbound $5.00

OLD-TIME SCHOOLS AND SCHOOL BOOKS, Clifton Johnson. Illustrations and rhymes from early primers, abundant quotations from early textbooks, many anecdotes of school life enliven this study of elementary schools from Puritans to middle 19th century. Introduction by Carl Withers. 234 illustrations. xxxiii + 381pp.
21031-6 Paperbound $2.50

VISUAL ILLUSIONS: THEIR CAUSES, CHARACTERISTICS, AND APPLICATIONS, Matthew Luckiesh. Thorough description and discussion of optical illusion, geometric and perspective, particularly; size and shape distortions, illusions of color, of motion; natural illusions; use of illusion in art and magic, industry, etc. Most useful today with op art, also for classical art. Scores of effects illustrated. Introduction by William H. Ittleson. 100 illustrations. xxi + 252pp.

21530-X Paperbound $2.00

A HANDBOOK OF ANATOMY FOR ART STUDENTS, Arthur Thomson. Thorough, virtually exhaustive coverage of skeletal structure, musculature, etc. Full text, supplemented by anatomical diagrams and drawings and by photographs of undraped figures. Unique in its comparison of male and female forms, pointing out differences of contour, texture, form. 211 figures, 40 drawings, 86 photographs. xx + 459pp. 5⅜ x 8⅜.

21163-0 Paperbound $3.50

150 MASTERPIECES OF DRAWING, Selected by Anthony Toney. Full page reproductions of drawings from the early 16th to the end of the 18th century, all beautifully reproduced: Rembrandt, Michelangelo, Dürer, Fragonard, Urs, Graf, Wouwerman, many others. First-rate browsing book, model book for artists. xviii + 150pp. 8⅜ x 11¼.

21032-4 Paperbound $2.50

THE LATER WORK OF AUBREY BEARDSLEY, Aubrey Beardsley. Exotic, erotic, ironic masterpieces in full maturity: Comedy Ballet, Venus and Tannhauser, Pierrot, Lysistrata, Rape of the Lock, Savoy material, Ali Baba, Volpone, etc. This material revolutionized the art world, and is still powerful, fresh, brilliant. With *The Early Work*, all Beardsley's finest work. 174 plates, 2 in color. xiv + 176pp. 8⅛ x 11.

21817-1 Paperbound $3.00

DRAWINGS OF REMBRANDT, Rembrandt van Rijn. Complete reproduction of fabulously rare edition by Lippmann and Hofstede de Groot, completely reedited, updated, improved by Prof. Seymour Slive, Fogg Museum. Portraits, Biblical sketches, landscapes, Oriental types, nudes, episodes from classical mythology—All Rembrandt's fertile genius. Also selection of drawings by his pupils and followers. "Stunning volumes," *Saturday Review*. 550 illustrations. lxxviii + 552pp. 9⅛ x 12¼.

21485-0, 21486-9 Two volumes, Paperbound $7.00

THE DISASTERS OF WAR, Francisco Goya. One of the masterpieces of Western civilization—83 etchings that record Goya's shattering, bitter reaction to the Napoleonic war that swept through Spain after the insurrection of 1808 and to war in general. Reprint of the first edition, with three additional plates from Boston's Museum of Fine Arts. All plates facsimile size. Introduction by Philip Hofer, Fogg Museum. v + 97pp. 9⅜ x 8¼.

21872-4 Paperbound $2.00

GRAPHIC WORKS OF ODILON REDON. Largest collection of Redon's graphic works ever assembled: 172 lithographs, 28 etchings and engravings, 9 drawings. These include some of his most famous works. All the plates from *Odilon Redon: oeuvre graphique complet,* plus additional plates. New introduction and caption translations by Alfred Werner. 209 illustrations. xxvii + 209pp. 9⅛ x 12¼.

21966-8 Paperbound $4.00

JIM WHITEWOLF: THE LIFE OF A KIOWA APACHE INDIAN, Charles S. Brant, editor. Spans transition between native life and acculturation period, 1880 on. Kiowa culture, personal life pattern, religion and the supernatural, the Ghost Dance, breakdown in the White Man's world, similar material. 1 map. xii + 144pp.
22015-X Paperbound $1.75

THE NATIVE TRIBES OF CENTRAL AUSTRALIA, Baldwin Spencer and F. J. Gillen. Basic book in anthropology, devoted to full coverage of the Arunta and Warramunga tribes; the source for knowledge about kinship systems, material and social culture, religion, etc. Still unsurpassed. 121 photographs, 89 drawings. xviii + 669pp.
21775-2 Paperbound $5.00

MALAY MAGIC, Walter W. Skeat. Classic (1900); still the definitive work on the folklore and popular religion of the Malay peninsula. Describes marriage rites, birth spirits and ceremonies, medicine, dances, games, war and weapons, etc. Extensive quotes from original sources, many magic charms translated into English. 35 illustrations. Preface by Charles Otto Blagden. xxiv + 685pp.
21760-4 Paperbound $4.00

HEAVENS ON EARTH: UTOPIAN COMMUNITIES IN AMERICA, 1680-1880, Mark Holloway. The finest nontechnical account of American utopias, from the early Woman in the Wilderness, Ephrata, Rappites to the enormous mid 19th-century efflorescence; Shakers, New Harmony, Equity Stores, Fourier's Phalanxes, Oneida, Amana, Fruitlands, etc. "Entertaining and very instructive." *Times Literary Supplement.* 15 illustrations. 246pp.
21593-8 Paperbound $2.00

LONDON LABOUR AND THE LONDON POOR, Henry Mayhew. Earliest (c. 1850) sociological study in English, describing myriad subcultures of London poor. Particularly remarkable for the thousands of pages of direct testimony taken from the lips of London prostitutes, thieves, beggars, street sellers, chimney-sweepers, street-musicians, "mudlarks," "pure-finders," rag-gatherers, "running-patterers," dock laborers, cab-men, and hundreds of others, quoted directly in this massive work. An extraordinarily vital picture of London emerges. 110 illustrations. Total of lxxvi + 1951pp. 6⅝ x 10.
21934-8, 21935-6, 21936-4, 21937-2 Four volumes, Paperbound $14.00

HISTORY OF THE LATER ROMAN EMPIRE, J. B. Bury. Eloquent, detailed reconstruction of Western and Byzantine Roman Empire by a major historian, from the death of Theodosius I (395 A.D.) to the death of Justinian (565). Extensive quotations from contemporary sources; full coverage of important Roman and foreign figures of the time. xxxiv + 965pp. 21829-5 Record, book, album. Monaural. $3.50

AN INTELLECTUAL AND CULTURAL HISTORY OF THE WESTERN WORLD, Harry Elmer Barnes. Monumental study, tracing the development of the accomplishments that make up human culture. Every aspect of man's achievement surveyed from its origins in the Paleolithic to the present day (1964); social structures, ideas, economic systems, art, literature, technology, mathematics, the sciences, medicine, religion, jurisprudence, etc. Evaluations of the contributions of scores of great men. 1964 edition, revised and edited by scholars in the many fields represented. Total of xxix + 1381pp. 21275-0, 21276-9, 21277-7 Three volumes, Paperbound $7.75

MATHEMATICAL PUZZLES FOR BEGINNERS AND ENTHUSIASTS, Geoffrey Mott-Smith. 189 puzzles from easy to difficult—involving arithmetic, logic, algebra, properties of digits, probability, etc.—for enjoyment and mental stimulus. Explanation of mathematical principles behind the puzzles. 135 illustrations. viii + 248pp.
20198-8 Paperbound $1.75

PAPER FOLDING FOR BEGINNERS, William D. Murray and Francis J. Rigney. Easiest book on the market, clearest instructions on making interesting, beautiful origami. Sail boats, cups, roosters, frogs that move legs, bonbon boxes, standing birds, etc. 40 projects; more than 275 diagrams and photographs. 94pp.
20713-7 Paperbound $1.00

TRICKS AND GAMES ON THE POOL TABLE, Fred Herrmann. 79 tricks and games— some solitaires, some for two or more players, some competitive games—to entertain you between formal games. Mystifying shots and throws, unusual caroms, tricks involving such props as cork, coins, a hat, etc. Formerly *Fun on the Pool Table.* 77 figures. 95pp.
21814-7 Paperbound $1.00

HAND SHADOWS TO BE THROWN UPON THE WALL: A SERIES OF NOVEL AND AMUSING FIGURES FORMED BY THE HAND, Henry Bursill. Delightful picturebook from great-grandfather's day shows how to make 18 different hand shadows: a bird that flies, duck that quacks, dog that wags his tail, camel, goose, deer, boy, turtle, etc. Only book of its sort. vi + 33pp. 6½ x 9¼.   21779-5 Paperbound $1.00

WHITTLING AND WOODCARVING, E. J. Tangerman. 18th printing of best book on market. "If you can cut a potato you can carve" toys and puzzles, chains, chessmen, caricatures, masks, frames, woodcut blocks, surface patterns, much more. Information on tools, woods, techniques. Also goes into serious wood sculpture from Middle Ages to present, East and West. 464 photos, figures. x + 293pp.
20965-2 Paperbound $2.00

HISTORY OF PHILOSOPHY, Julián Marias. Possibly the clearest, most easily followed, best planned, most useful one-volume history of philosophy on the market; neither skimpy nor overfull. Full details on system of every major philosopher and dozens of less important thinkers from pre-Socratics up to Existentialism and later. Strong on many European figures usually omitted. Has gone through dozens of editions in Europe. 1966 edition, translated by Stanley Appelbaum and Clarence Strowbridge. xviii + 505pp.   21739-6 Paperbound $3.00

YOGA: A SCIENTIFIC EVALUATION, Kovoor T. Behanan. Scientific but non-technical study of physiological results of yoga exercises; done under auspices of Yale U. Relations to Indian thought, to psychoanalysis, etc. 16 photos. xxiii + 270pp.
20505-3 Paperbound $2.50

Prices subject to change without notice.
Available at your book dealer or write for free catalogue to Dept. GI, Dover Publications, Inc., 180 Varick St., N. Y., N. Y. 10014. Dover publishes more than 150 books each year on science, elementary and advanced mathematics, biology, music, art, literary history, social sciences and other areas.